Vermicompost in Sustainable Crop Production

Vermicompost in Sustainable Crop Production

Guest Editor

Gederts Ievinsh

MDPI

Basel • Beijing • Wuhan • Barcelona • Belgrade • Novi Sad • Cluj • Manchester

Guest Editor
Gederts Ievinsh
Department of Plant
Physiology
University of Latvia
Rīga
Latvia

Editorial Office
MDPI AG
Grosspeteranlage 5
4052 Basel, Switzerland

This is a reprint of the Special Issue, published open access by the journal *Agriculture* (ISSN 2077-0472), freely accessible at: https://www.mdpi.com/journal/agriculture/special_issues/vermicompost_crop_production.

For citation purposes, cite each article independently as indicated on the article page online and as indicated below:

Lastname, A.A.; Lastname, B.B. Article Title. *Journal Name* **Year**, *Volume Number*, Page Range.

ISBN 978-3-7258-4953-6 (Hbk)
ISBN 978-3-7258-4954-3 (PDF)
https://doi.org/10.3390/books978-3-7258-4954-3

Contents

Preface

The aim of this reprint is to provide up-to-date information on recent scientific studies involving vermicompost. Vermicompost is organic fertilizer produced by earthworms and their symbiotic microorganisms. Vermicompost contains plant-available soluble minerals and organic matter, providing essential nutrients for plant growth. It enhances soil mineral nutrient availability and supports gradual nutrient release through microbial activity. In addition, it contains hormonelike substances (e.g., auxins, cytokinins, and gibberellins) and humic substances that stimulate root elongation, lateral root formation, and overall plant growth. Additional benefits of vermicompost application include improvement of soil structure, water-holding capacity, and microbial diversity, contributing to long-term soil health and sustainability. Application of vermicompost enhances plant resistance to unfavorable environmental factors, pathogens, and herbivores. These benefits make vermicompost a valuable organic fertilizer for both organic and conventional farming systems. The information included in this reprint will therefore be of particular benefit for both agricultural and environmental scientists and agricultural practitioners.

Gederts Ievinsh
Guest Editor

agriculture

MDPI

Article

Substrate-Dependent Effect of Vermicompost on Yield and Physiological Indices of Container-Grown *Dracocephalum moldavica* Plants

Alise Ose, Una Andersone-Ozola and Gederts Ievinsh *

Department of Plant Physiology, Faculty of Biology, University of Latvia, 1 Jelgavas Str., LV-1004 Riga, Latvia; alise.ose7@gmail.com (A.O.); una.andersone-ozola@lu.lv (U.A.-O.)

* Correspondence: gederts.ievins@lu.lv

Abstract: The development of sustainable plant production systems involves a search for different alternatives to chemical fertilizers. The aim of the present study is to compare growth and physiological effects of vermicompost on *Dracocephalum moldavica* plants in controlled conditions, using two types of commercially available substrates. The intention is to determine whether nondestructively measured photosynthesis-related parameters are useful for monitoring the physiological status of plants. The plants were cultivated in two base substrates without or with the addition of mineral fertilizer, as well as an amendment with vermicompost at a 20% or 30% rate in the conditions of an automated greenhouse. The biomass accumulation for control plants of *D. moldavica* was identical in peat substrate and commercial garden soil. The average growth increase by mineral fertilizer was 25% for *D. moldavica* plants grown in peat and 15% for plants grown in soil. Substrate amendment with 20% vermicompost resulted in an 114% average increase in biomass for plants grown in peat and a 98% average increase for plants grown in soil, but for plants at 30% the amendment rate increase was 148% and 68%, for peat and soil, respectively. Consequently, the addition of an identical amount of vermicompost resulted in a poorer growth response of plants in commercial garden soil as a substrate in comparison to peat, but an increase in the amendment rate from 20% to 30% resulted in some growth inhibition for these plants. Chlorophyll concentration was positively affected by the vermicompost amendment in a concentration-dependent manner, but this effect during a cultivation period appeared relatively late. Large differences were found between the three groups of fluorescence-derived parameters, with variable levels of predictability with respect to the differences in plant yield due to the pronounced variation in correlation through time. It is concluded that the incorporation of vermicompost for the cultivation of *D. moldavica,* even in substrate mixes with relatively high and balanced composition of plant-available nutrients, benefits plant growth, physiological status and biomass yield, but it is necessary to explore interactions between vermicompost and other substrates leading to possible changes in quality-related characteristics of vermicompost in substrate mixes.

Keywords: chlorophyll; chlorophyll *a* fluorescence; *Dracocephalum moldavica*; growth; sustainable production; vermicompost

1. Introduction

The use of renewable resources for nutrient management is an important aspect of sustainable agricultural and horticultural practices. Therefore, the development of sustainable plant production systems involves the search for different alternatives to chemical fertilizers [1]. One extremely promising direction in this respect is related to the application of vermicompost, which is a type of organic fertilizer produced by the concerted action of earthworms and their symbiotic microorganisms [2]. The application of vermicompost leads to increased soil sustainability, due to the enhancement of organic matter content and microbial diversity, as well as to the improvement of the physical properties of soil [3,4]. In addition, there are direct benefits for crop plants from the use of vermicompost. The physiological effects of vermicompost on plants has been recently reviewed and it was concluded that in conventional farming systems, vermicompost can substitute chemical fertilizers due to the significant concentration of plant-available mineral nutrients, while additional benefits have been associated with the presence of plant growth-stimulating substances and adaptogenic activity [5].

Traditionally, vermicomposts are produced from cattle manure with the addition of organic carbon-rich biomass, such as grass or straw [6], but it is possible to use different organic waste materials and byproducts for its production. The possibility for successful application of different types of vermicompost for the production of various crop species has been shown for vermicomposts produced from agricultural waste [7], byproducts of food production or food waste [8,9], urban organic waste [10], and composted sewage sludge [11].

Recently, several reviews and meta-analyses have summarized the beneficial effects of vermicompost-use in different farming systems, which also indicates directions for further studies of critical importance [5,12–15]. Organic agriculture and horticulture are especially important targets for vermicompost application, in a view of providing balanced nutrition for crop needs and soil sustainability through enhanced microbiological activity [16,17]. Another especially promising direction in horticulture is the inclusion of vermicompost as a major component in substrate mixes for the design of alternative plant-growing media to replace peat-based mixes [18,19]. While previous studies in this direction have produced promising results, physicochemical interactions between vermicompost and other substrate components, and their possible effect on the beneficial activity of vermicompost, have not been experimentally assessed.

Moldavian dragonhead or Moldavian balm (*Dracocephalum moldavica* L.) is a species of Lamiaceae family with a characteristic high content of essential oil in above-ground parts and fatty oil-producing seeds. The agrobiological characteristics of *D. moldavica* have been reviewed recently [20], and it was described as medicinal, spice, and nectar plant, with a high potential for the use of seeds in different food applications. Therefore, there is an increasing interest in the development of optimum cultivation technologies for this crop species.

The different aspects of *D. moldavica* fertilization in field conditions have been studied, including the effects of compost [21,22], compost and farmyard manure [23], forms of nitrogen [24], arbuscular mycorrhiza [25], and vermicompost [26], with a general aim to improve yield and quality of the plant while focusing on soil sustainability. Most importantly, the use of different organic fertilizers clearly promotes plant growth by providing additional plant-available nutrients to the soil and improving the essential oil content. It needs to be emphasized that the generalization of this type of information to establish best agrotechnical practices for the cultivation of *D. moldavica* is difficult because of the high variability of results between individual studies, due to differences in agroecological conditions [20]. Studies in controlled conditions greatly eliminate the undesirable effects of environmental heterogeneity, providing more comparable results with higher generalization ability. Recently, we examined a possibility for the organic production of *D. moldavica* in controlled conditions, using compost and vermicompost as soil amendments, and concluded that the use of vermicompost is superior to that of compost, even with the same amount of plant-available nutrients [27].

The nondestructive evaluation of the physiological status of intact growing plants often involves a measurement of leaf chlorophyll content by chlorophyll meters as well as a chlorophyll *a* fluorescence analysis, which characterizes the photochemical activity of photosynthesis [28,29]. The analysis of fluorescence transient has been most widely used as a tool to monitor environmental stress responses in plants [30]. However, it can also be used to characterize changes in the general physiological status of plants, as related to their performance and environmental adaptation in the case of wild plants [28] or growth/yield relationships of crop species due to different agricultural practices [27]. In particular, several studies have shown a relationship between the photochemistry of photosynthesis and the status of plant mineral nutrition [31–34], indicating that chlorophyll *a* fluorescence analysis can be also used to predict the physiological status of crops in studies aiming to assess the effects of organic fertilizers.

The aim of the present study is to compare the growth and physiological effects of vermicompost on *D. moldavica* plants in controlled conditions, using two types of commercially available substrates. The second aim is to assess if nondestructively measured photosynthesis-related parameters, leaf chlorophyll concentration, and chlorophyll *a* fluorescence-related indices, are useful for monitoring the physiological status of plants under different regimes of substrate amendment with the organic fertilizer, vermicompost.

2. Materials and Methods

2.1. Plant Material, Substrates, and Vermicompost

The study was performed with *Dracocephalum moldavica* L. plants grown from seeds purchased from Kurzemes Sēklas (Talsi, Latvia). Peat substrate KKS-1 (Laflora, Jelgava District, Latvia), commercial garden soil (Biolan, Eura, Finland), and vermicompost (Eko Zeme, Bauska District, Latvia) were purchased from local suppliers. The peat substrate contained fertilizer PG-Mix (15-10-20, 1 kg m^{-3}), limestone (6 kg m^{-3}), dolomite (1.8 kg m^{-3}) and a wetting agent, Instant (0.3 L m^{-3}). The garden soil contained mineral fertilizer (12-14-24, 1 kg m^{-3}) and dolomite (4 kg m^{-3}). Vermicompost was produced from composted cow manure and grass biomass.

The analysis of plant-available mineral nutrient concentration in the substrates and vermicompost was performed in a certified agrochemical laboratory (Laboratory of Plant Mineral Nutrition, Institute of Biology, University of Latvia). The results shown in Table 1 indicate that peat substrate had higher electrical conductivity (EC) in comparison to commercial garden soil, pointing to a higher concentration of soluble ions. Both substrates had a similar pH and concentration of plant-available K, Mg, Mn, Zn, and Cu. However, the N and P concentration was higher in peat substrate, but Ca, S, and Fe were higher in garden soil. In respect to the optimum element concentrations for cultivated plants, the peat substrate showed optimum or above-optimum levels, while a deficiency of N was evident for garden soil. Both substrates were extremely rich in Ca and Mg. Opposite to this, vermicompost was very good source of plant-available N, P, K, Mg, Mn, and Zn.

Table 1. Concentration of plant-available mineral nutrients (mg L^{-1} dry substrate) in substrate and vermicompost samples used in the present study in comparison to approximate optimum concentrations in substrate for cultivated plants [35].

Nutrient	Peat Substrate	Commercial Garden Soil	Vermicompost	Optimum for Cultivated Plants
N	120	38	670	120
P	95	55	1997	60
K	225	240	8300	150
Ca	3125	6450	7850	800
Mg	595	550	3600	50
S	50	110	275	50
Fe	90	510	365	30
Mn	7.5	11.5	105.0	1.5
Zn	2.0	2.5	40.0	1.0
Cu	2.45	2.15	4.70	0.50
Mo	0.24	0.06	0.03	0.02
B	0.4	1.3	2.3	0.2
pH_{KCl}	6.39	6.28	8.03	n.a.
EC (mS m^{-1})	202	145	2205	n.a.

Elements were analyzed in 1 M HCl substrate extract.

2.2. Plant Establishment, Treatments, and Cultivation Conditions

Seeds of *D. moldavica* were sown in heated (60 °C, 24 h) commercial garden soil (Biolan, Eura, UK) in 1 L plastic plant tissue culture containers, and closed and kept for two weeks in a growth cabinet (light/dark period of 16/8 h, photosynthetically active radiation with a photon flux density 100 µmol m^{-2} s^{-1}, day/night temperature 15/20 °C). After that, the developed seedlings were individually transplanted to 200 mL plastic containers filled with commercial garden soil (Biolan, Eura, Finland). Containers with plants were placed in 48 L plastic boxes, closed with lids, and located in an experimental greenhouse with automatic control system (HortiMaX, Maasdijk, The Netherlands). Additional light was supplemented by Master SON-TPIA Green Power CG T 400 W (Philips, Amsterdam, Netherlands) and Powerstar HQI-BT 400 W/D PRO (Osram, Munich, Germany) lamps (380 µmol m^{-2} s^{-1} at the plant level) for a 16 h photoperiod, with day/night temperature 25/16 °C, and a relative air humidity of 60% to 70%. The plants were watered with deionized water. The boxes were periodically ventilated for the acclimation of the seedlings to greenhouse conditions. After two weeks, when the plants reached 5 to 10 cm height and developed four true leaves, they were used for the experiment.

The plants were transplanted to 1.2 L plastic containers filled with different substrate mixes. In total, eight treatments in five replicates were established, including two controls. Two commercial substrates, peat substrate KKS-1 (Laflora, Jelgava District, Latvia) and commercial compost-based garden soil (Biolan, Eura, Finland) were used as an alternative basis for the formation of substrate mixes. Peat substrates without the additives and garden soil without additives were used as the two controls. For two mineral fertilizer treatments, plants growing in peat substrate without additives or garden soil without additives received mineral fertilizer biweekly, starting from the third week after transplantation, using 0.075% Yara Tera Kristalon Blue fertilizer (19-6-20+MgO+micro; Yara International, Oslo, Norway), 200 mL per container. Four treatments with vermicompost were made by adding 20% or 30% (*v*/*v*) vermicompost to both peat substrate or garden soil. Individual containers were randomly arranged on the greenhouse bench and their location was changed biweekly. Conditions in the greenhouse were the same as described above. The substrate water content was monitored with a HH2 moisture meter equipped with WET-2 sensor (Delta-T Devices, Burwell, UK) and kept at 50% to 60% using deionized water. The plants were cultivated for eight weeks.

Substrate electrical conductivity (EC) and pH were measured 1 week and 8 weeks after the start of the experiment. For EC, the HH2 moisture meter equipped with a WET-2 sensor (Delta-T Devices, Burwell, UK) was used. Substrate pH was measured using a pH meter pH 3000 (STEP Systems, Nürnberg, Germany). For every container, four separate measurements on all sides of the container were performed for both measurements.

2.3. Measurement of Photosynthesis-Related Parameters

Measurements of photosynthesis-related parameters were performed weekly. The leaf chlorophyll concentration was measured using a chlorophyll meter CCM-300 (Opti-Sciences, Hudson, NH, USA). Four fully grown and actively photosynthesizing leaves from three randomly selected plants per treatment were measured. Chlorophyll *a* fluorescence was measured in four leaves dark-adapted for at least 20 min by the Handy PEA fluorometer (Hansatech Instruments, King's Lynn, UK) on each of the three plants per treatment.

Fluorescence data analysis was performed by PEA Plus software (Hansatech Instruments, King's Lynn, UK). A number of parameters derived from the fast fluorescence induction curve were used for the analysis [36,37]. F_v/F_m, calculated as $(F_m - F_0)/F_m$, and represents the maximum quantum efficiency of photosystem II (PSII), indicating a probability that a trapped photon will perform a further photochemical energy transfer. F_v/F_0, calculated as $(F_m - F_0)/F_0$, is considered to reflect an instant photochemical activity at the donor side of PSII. Performance Index (PI) represents the multiparametric and multitypal entity, used as relative indication of sample vitality, and can have different types of expression. Thus, PI_{inst} combines three function-related (trapping of absorbed exciton, electron transport between the photosystems, and reduction of end-electron acceptors) parameters. PI_{abs} indicates the functional activity of PSII related to the energy absorbed, as in addition to the three previous parameters, it also includes a structural parameter, the amount of chlorophyll per reaction center of chlorophyll. PI_{total} includes information on the status of both PSII and photosystem I (PSI), in addition to characterizing the electron flow between the two systems, which is also on an absorption basis.

2.4. Morphological Measurements

At the termination of the experiment, individual plants were cut, plant height was measured, the number of branches was counted, and the leaves were divided from the stems and weighed separately. Plant tissues were dried in an oven at 60 °C for 72 h and the dry mass was measured. The water content in plant leaves and stems was calculated in g H_2O per g dry mass.

2.5. Data Analysis

The measurement results were analyzed and the graphs were made by KaleidaGraph (v. 4.1, Synergy Software, Reading, PA, USA). The statistical significance of the differences of individual parameters between treatments was evaluated by one-way ANOVA minimum significant difference tests using a Microsoft Excel spreadsheet (www.biostathandbook.com/anova.xls, accessed on 15 July 2021) [38].

3. Results

3.1. Effect of Substrate Type, Fertilizer, and Vermicompost on EC and pH

The substrate electrical conductivity (EC) was measured weekly throughout the experimental period to monitor changes in soluble salt concentration (Figure 1). A characteristic decrease of EC values during the cultivation of *D. moldavica* reflected the uptake of mineral nutrients by the plants. The initial EC in peat substrate was 8% higher than that in commercial garden soil, and the difference was statistically significant. Moreover, the addition of equal amounts of vermicompost resulted in a higher increase in substrate EC in the case of peat. Thus, EC increased by 268 and 411 mS m^{-1} for a 20% and 30% amendment, respectively, in peat substrate, and only by 242 and 324 mS m^{-1} for a 20% and 30% amendment, respectively, in commercial garden soil. Similarly, a decrease of EC during plant cultivation was more pronounced in the case of peat (356 and 444 mS m^{-1} for a 20% and 30% amendment, respectively) in comparison to that in soil (292 and 345 mS m^{-1} for 20% and 30% amendment, respectively). However, a decrease of substrate EC for control plants was identical in peat and soil (by 145 mS m^{-1}).

Initial pH in peat substrate was significantly higher than that in commercial garden soil (Figure 2A). An amendment with vermicompost significantly increased the pH in peat (by 4% and 6%, for a 20% and 30% amendment rate, respectively), but the increase in soil by the vermicompost amendment was more pronounced (by 16% and 17%, for a 20% and 30% amendment rate, respectively). The substrate pH further increased in all treatments during plant cultivation, but initial differences between peat and soil treatments remained significant (Figure 2B).

3.2. Effect of Substrate Type, Fertilizer, and Vermicompost on Morphological Parameters

As for morphological appearance, *D. moldavica* plants grown in vermicompost-amended substrate had denser and greener foliage, in comparison to the control and mineral fertilizer-treated plants (Figure 3). However, the growth of control plants in peat substrate or commercial garden soil was identical (Figures 4 and 5), except that a fresh mass of stems was 16% lower in commercial garden soil in comparison to that in peat substrate (Figure 4B). The addition of mineral fertilizer increased both the fresh and dry mass of the leaves and stems, and this effect was significant for all parameters for peat-grown plants, but only for the fresh mass of leaves for soil-grown plants. The average growth increase by mineral fertilizer was 25% for *D. moldavica* plants grown in peat and 15% for plants grown in soil. The substrate amendment with 20% vermicompost resulted in a 114% average increase in biomass for plants grown in peat and a 98% average increase for plants grown in soil; however, for plants at a 30% amendment rate, the increase was 148% and 68% for peat and soil, respectively. Thus, for *D. moldavica* plants grown in peat, an amendment with 30% vermicompost resulted in an additional statistically significant increase of plant biomass, but that was not the case for plants grown in soil, with a significant decrease in the dry mass of both the leaves and stems in comparison to a 20% vermicompost amendment. As a result, at the highest vermicompost amendment rate, the plants grown in soil had a 47% and 30% lower fresh mass of the leaves and stems, as well as a 37% and 34% lower dry mass of the leaves and stems, in comparison to peat-grown plants.

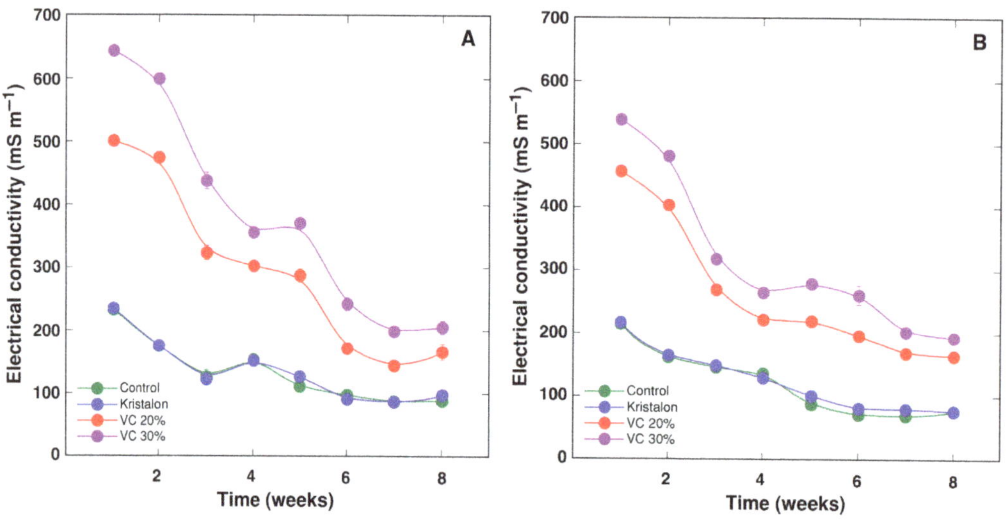

Figure 1. Changes of substrate electrical conductivity as affected by mineral fertilizer (Kristalon) and amendment with vermicompost (VC) at a 20% and 30% (v/v) rate during cultivation of *Dracocephalum moldavica* plants in peat substrate (**A**) or commercial garden soil (**B**). Results are means ± SE from five replicates for each treatment, four measurements per replicate.

The water content of leaves showed only negligible differences between treatments (Figure 6A), but the water content of the stems tended to be lower for plants grown in soil (Figure 6B). Similarly, plant height tended to be lower for soil-grown plants in all treatments, with significant differences for both vermicompost amendment rates (Figure 7A). In addition, the number of branches increased in vermicompost-amended plants (Figure 7B).

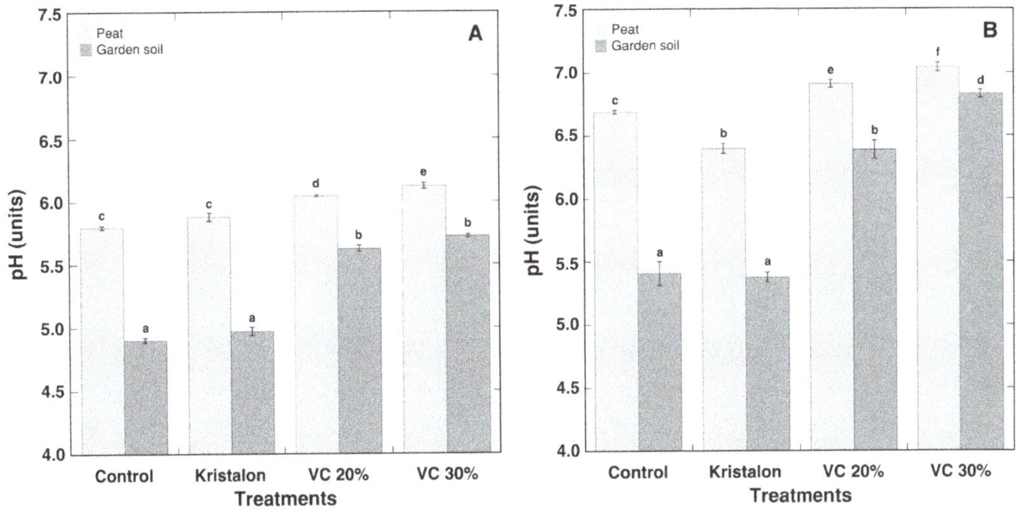

Figure 2. Effect of mineral fertilizer (Kristalon) and amendment with vermicompost (VC) at a 20% and 30% (v/v) rate on substrate pH for *Dracocephalum moldavica* plants grown in peat substrate or commercial garden soil for 1 week (**A**) and 8 weeks (**B**) after the start of the experiment. Results are means ± SE from five replicates for each treatment, four measurements per replicate. Means with identical letters are not considered statistically significantly different ($p < 0.05$).

Figure 3. Morphology of the typical average *Dracocephalum moldavica* grown in different substrates for 8 weeks. P, peat substrate; K, mineral fertilizer Kristalon; VC20%, 20% (v/v) vermicompost; VC30%, 30% (v/v) vermicompost; S, commercial garden soil.

3.3. *Effect of Substrate Type, Fertilizer, and Vermicompost on Photosynthesis-Related Parameters*

Leaf chlorophyll concentrations did not show any significant differences between treatments during four weeks of cultivation, where the concentration remained relatively stable (for plants in peat substrate) or showed an increasing trend (for plants in commercial garden soil) (Figure 8). Next, chlorophyll concentration decreased for plants in all treatments; however, the rate of decrease differed between the treatments, with more stable chlorophyll content for vermicompost-amended plants. As a result, there was no overall significant effect of vermicompost amendment on the chlorophyll concentration in both substrates (Table 2).

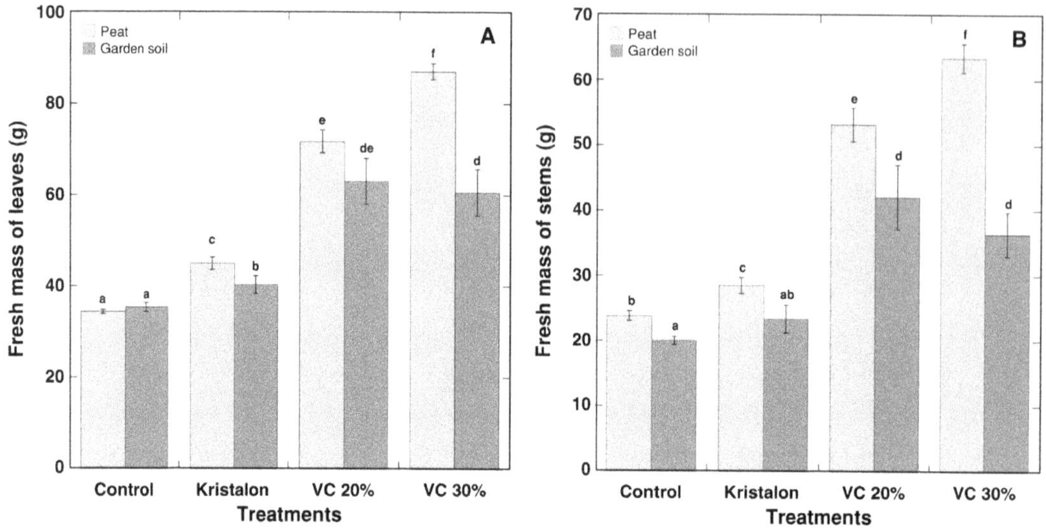

Figure 4. Effect of mineral fertilizer (Kristalon) and amendment with vermicompost (VC) at a 20% and 30% (v/v) rate on fresh mass of leaves (**A**) and stems with inflorescences (**B**) of *Dracocephalum moldavica* plants grown in peat substrate or commercial garden soil for 8 weeks. Results are means ± SE from five replicates for each treatment. Means with identical letters are not considered statistically significantly different ($p < 0.05$).

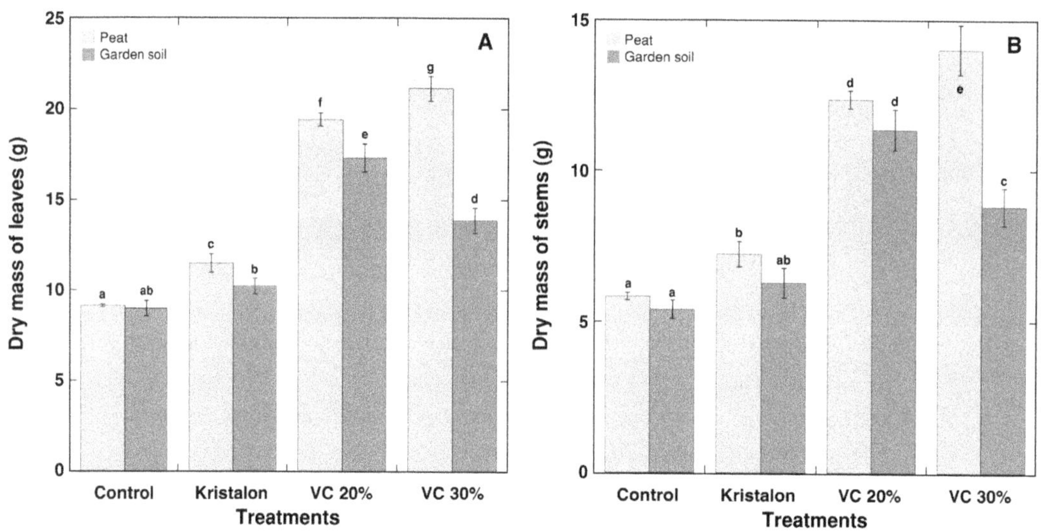

Figure 5. Effect of mineral fertilizer (Kristalon) and amendment with vermicompost (VC) at a 20% and 30% (v/v) rate on dry mass of leaves (**A**) and stems with inflorescences (**B**) of *Dracocephalum moldavica* plants grown in peat substrate or commercial garden soil for 8 weeks. Results are means ± SE from five replicates for each treatment. Means with identical letters are not considered statistically significantly different ($p < 0.05$).

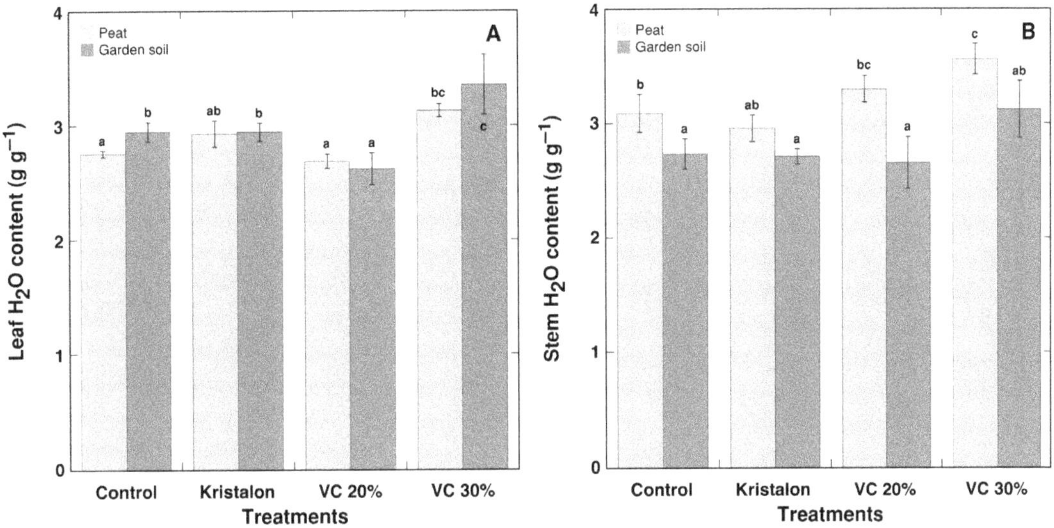

Figure 6. Effect of mineral fertilizer (Kristalon) and amendment with vermicompost (VC) at a 20% and 30% (v/v) rate on H_2O content in leaves (**A**) and stems with inflorescences (**B**) of *Dracocephalum moldavica* plants grown in peat substrate or commercial garden soil for 8 weeks. Results are means ± SE from five replicates for each treatment. Means with identical letters are not considered statistically significantly different ($p < 0.05$).

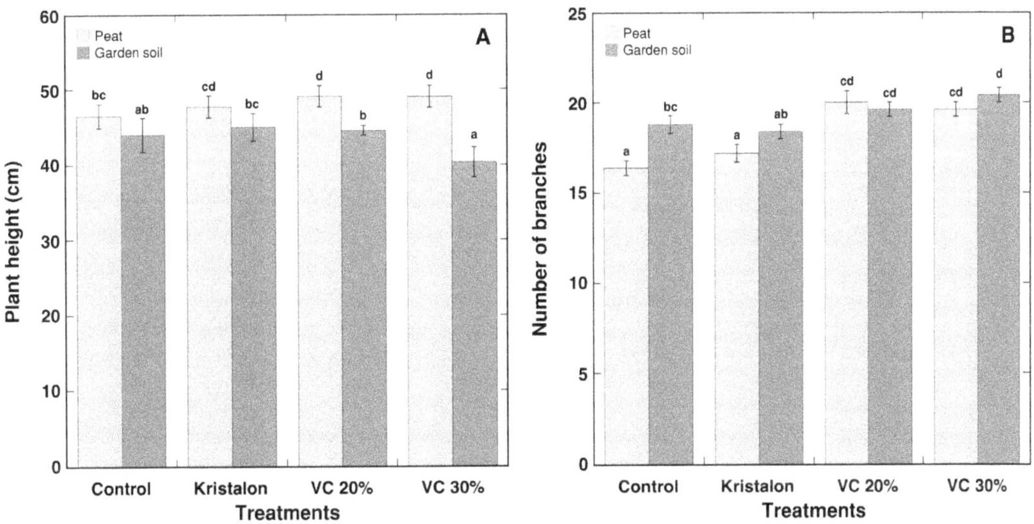

Figure 7. Effect of mineral fertilizer (Kristalon) and amendment with vermicompost (VC) at a 20% and 30% (v/v) rate on plant height (**A**) and number of branches (**B**) of *Dracocephalum moldavica* plants grown in peat substrate or commercial garden soil for 8 weeks. Results are means ± SE from five replicates for each treatment. Means with identical letters are not considered statistically significantly different ($p < 0.05$).

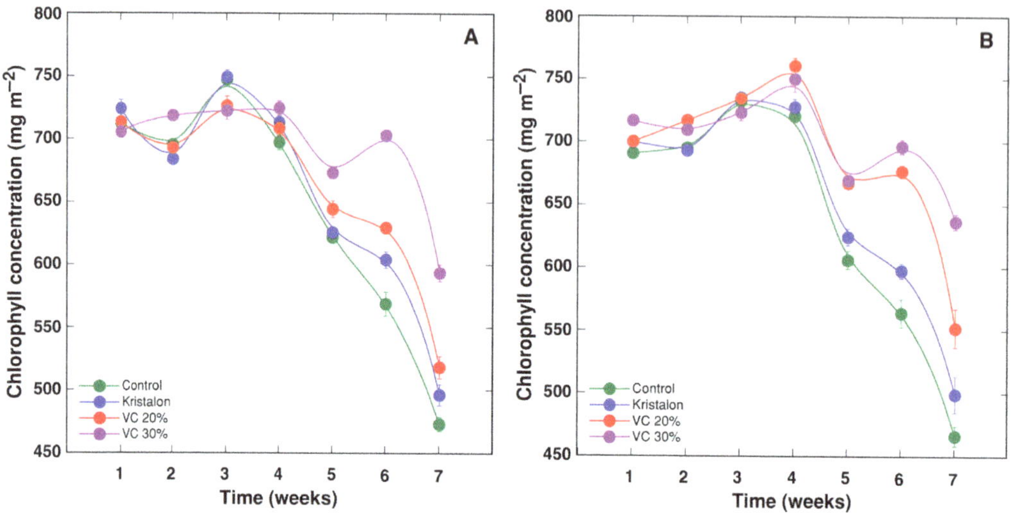

Figure 8. Changes of leaf chlorophyll concentration as affected by mineral fertilizer (Kristalon) and amendment with vermicompost (VC) at a 20% and 30% (v/v) rate in *Dracocephalum moldavica* plants grown in peat substrate (**A**) or commercial garden soil (**B**). Results are means ± SE from five replicates for each treatment, four measurements per replicate.

Table 2. Results of ANOVA analysis of physiological indices of *Dracocephalum moldavica* plants during cultivation in peat substrate and commercial garden soil with different amendments.

Parameter	Peat Substrate			Commercial Garden Soil		
	F	p	Significance Level	F	p	Significance Level
Chlorophyll Concentration	0.445	0.72275	n.s.	0.982	0.41759	n.s.
Fv/F_m	4.975	0.00797	<0.01	8.030	0.00071	<0.001
F_v/F_0	5.016	0.00769	<0.01	8.878	0.00038	<0.001
PI_{inst}	1.938	0.15030	n.s.	5.163	0.00677	<0.01
PI_{abs}	1.589	0.21815	n.s.	5.163	0.00678	<0.01
PI_{total}	0.303	0.82260	n.s.	0.473	0.70420	n.s.

Data analyzed are from Figures 8–12. n.s., not significant.

The chlorophyll *a* fluorescence fast induction curve in plant leaves was measured weekly during plant cultivation, and temporal changes of various parameters derived from the curve were compared in *D. moldavica* plants grown in different substrates (Figures 9–13). In general, the largest differences for F_v/F_m (Figure 9) and F_v/F_o (Figure 10) between treatments were evident at 4–6 weeks for plants grown in peat and at 5–6 weeks for plants grown in soil, with values for vermicompost-amended plants significantly higher than these for control or mineral-treated plants. Moreover, the statistically significant increase for plants amended with a 30% vermicompost in comparison to a 20% amendment was evident on week 6–7 for plants grown in peat (Figures 9A and 10A) and on week 4–7 for plants grown in soil (Figure 9B and Figure 10B). According to the ANOVA analysis, the effect of treatment was significant for both F_v/F_m and F_v/F_0, but the effect in the case of plants grown in soil was at a higher significance level than that of plants grown in peat (Table 2).

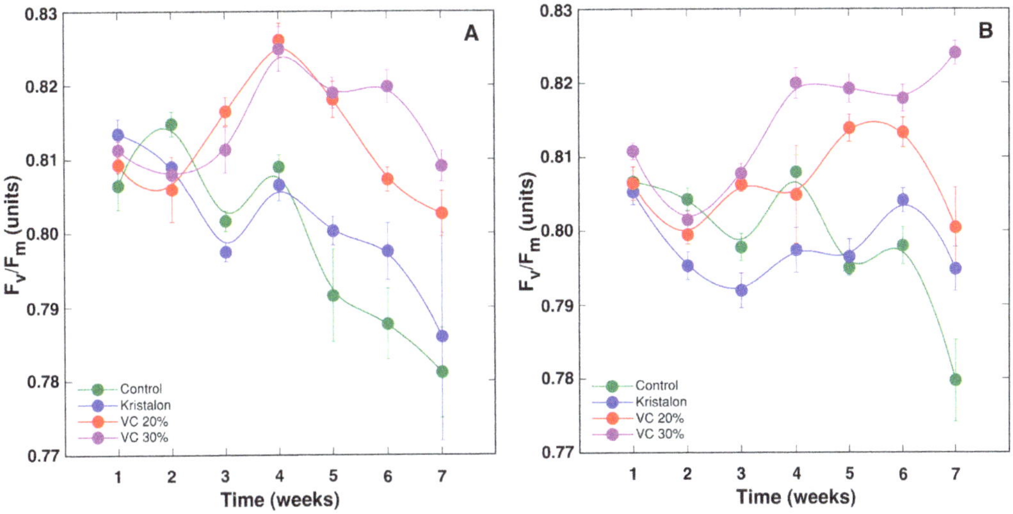

Figure 9. Changes of chlorophyll *a* fluorescence parameter F_v/F_m as affected by mineral fertilizer (Kristalon) and amendment with vermicompost (VC) at a 20% and 30% (*v*/*v*) rate in *Dracocephalum moldavica* plants grown in peat substrate (**A**) or commercial garden soil (**B**). Results are means ± SE from five replicates for each treatment, three measurements per replicate.

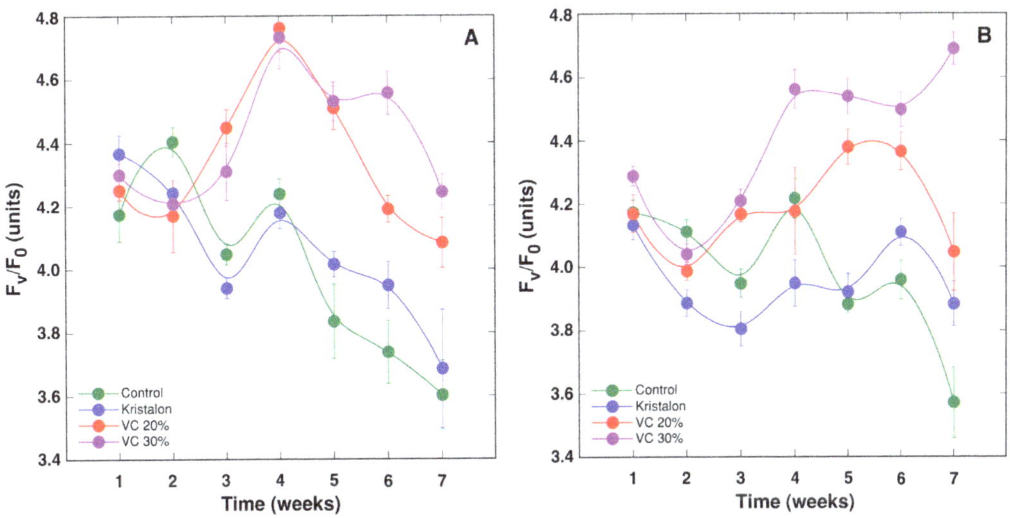

Figure 10. Changes of chlorophyll *a* fluorescence parameter F_v/F_0 as affected by mineral fertilizer (Kristalon) and amendment with vermicompost (VC) at a 20% and 30% (*v*/*v*) rate in *Dracocephalum moldavica* plants grown in peat substrate (**A**) or commercial garden soil (**B**). Results are means ± SE from five replicates for each treatment, three measurements per replicate.

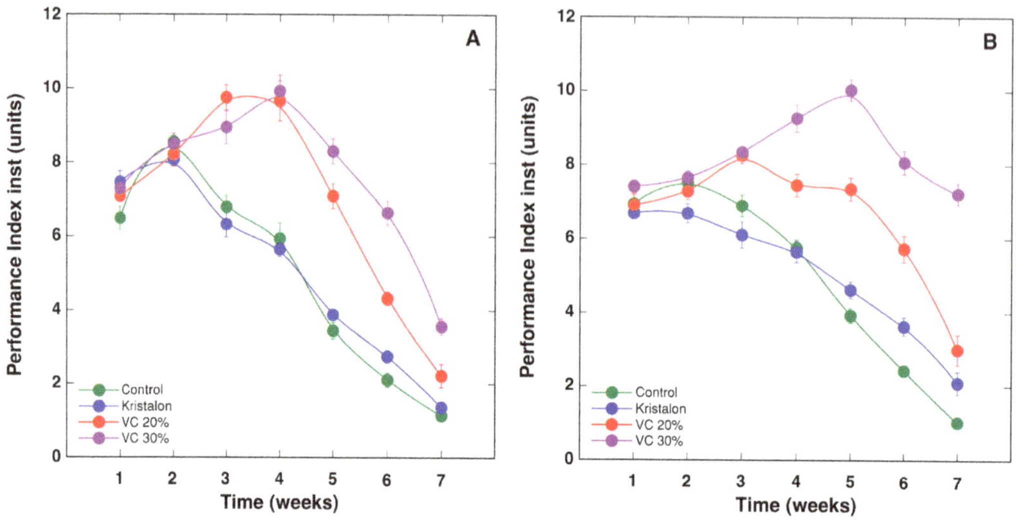

Figure 11. Changes of chlorophyll *a* fluorescence parameter PI_{inst}, as affected by mineral fertilizer (Kristalon) and amendment with vermicompost (VC) at a 20% and 30% (v/v) rate in *Dracocephalum moldavica* plants grown in peat substrate (**A**) or commercial garden soil (**B**). Results are means ± SE from five replicates for each treatment, three measurements per replicate.

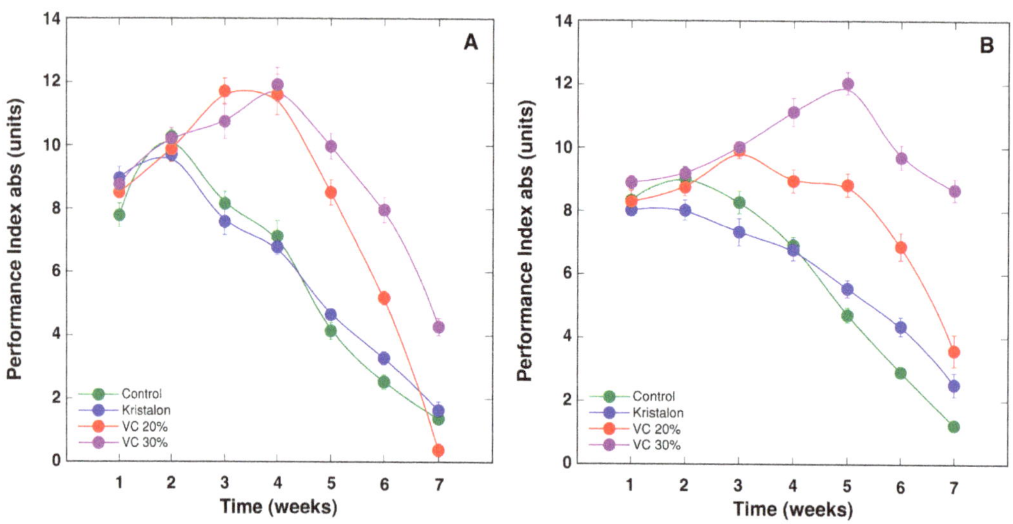

Figure 12. Changes of chlorophyll *a* fluorescence parameter PI_{abs}, as affected by mineral fertilizer (Kristalon) and amendment with vermicompost (VC) at a 20% and 30% (v/v) rate in *Dracocephalum moldavica* plants grown in peat substrate (**A**) or commercial garden soil (**B**). Results are means ± SE from five replicates for each treatment, three measurements per replicate.

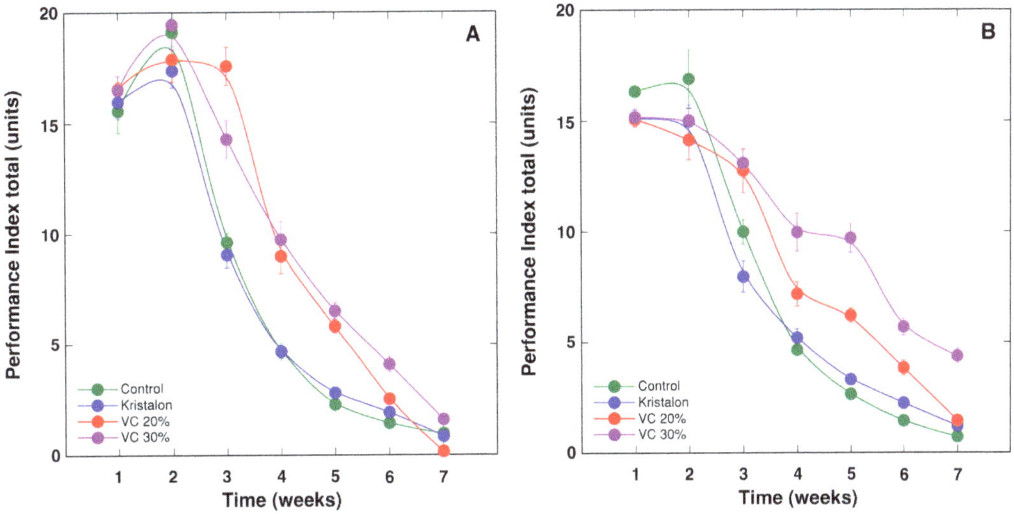

Figure 13. Changes of chlorophyll *a* fluorescence parameter P_{Itotal}, as affected by mineral fertilizer (Kristalon) and amendment with vermicompost (VC) at a 20% and 30% (v/v) rate in *Dracocephalum moldavica* plants grown in peat substrate (**A**) or commercial garden soil (**B**). Results are means ± SE from five replicates for each treatment, three measurements per replicate.

Two other, more complex parameters of chlorophyll *a* fluorescence, PI_{inst} and PI_{abs}, showed a relatively similar trend during plant cultivation (Figures 11 and 12). For vermicompost-amended plants grown in peat, both parameters increased up to week 4, and for amended plants at the highest rate in soil, up to week 5. For peat-grown plants, differences in PI_{inst} and PI_{abs} between the plants at two vermicompost amendment rates were less pronounced than these for soil-grown plants. As a result, the overall differences between treatments were not statistically significant for plants grown in peat, and significant for plants grown in soil (Table 2).

The chlorophyll *a* fluorescence parameter PI_{total} showed a pronounced decrease with plant age for all treatments, but some differences were evident between them (Figure 13). For plants grown in peat, the vermicompost amendment resulted in significantly higher PI_{total} during weeks 3–5, but there were no statistically significant differences between plants at two amendment rates (Figure 13A). For plants grown in soil, the vermicompost amendment resulted in significantly higher PI_{total} during weeks 3–6, and plants at a 30% amendment rate had significantly higher parameter values during weeks 4–7 (Figure 13B). However, the overall effect of treatments was not significant (Table 2).

To further analyze the possible contribution of chlorophyll concentration and a particular chlorophyll *a* fluorescence parameter to plant yield, as affected by different cultivation substrates, a correlation between the value of each physiological indication at each particular week of cultivation and each of four yield parameters (fresh and dry mass of leaves and stems) was calculated (Figure 14). The highest correlation between leaf chlorophyll concentration and yield was during weeks 5–6, evidently reflecting a delay of the senescence of plants grown in the vermicompost-amended substrate. In contrast, all fluorescence-derived parameters showed a high correlation with the yield at week 3, with pronounced differences between the parameters during the further cultivation period. Both PI_{inst} and PI_{abs} were the parameters with the highest degree of correlation during weeks 3–4; the correlation with F_v/F_m and F_v/F_0 peaked at week 5, but the correlation with PI_{total} steadily decreased (Figure 14).

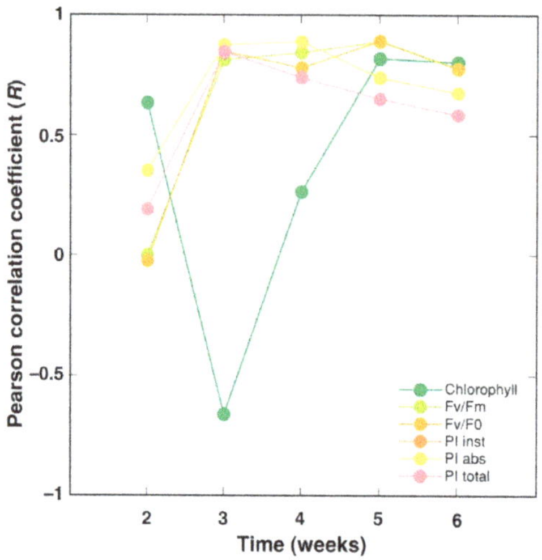

Figure 14. Changes of average correlation between physiological indices and yield parameters (fresh and dry mass of leaves and stems) of *Dracocephalum moldavica* plants during cultivation in peat substrate and commercial garden soil with different amendments. PI_{inst} has the same values as PI_{abs} and cannot be seen.

4. Discussion

Both compost [21–23,27] and vermicompost [26,27] have been verified as promising types of organic fertilizers for the cultivation of *D. moldavica*. Most importantly, the application of compost and vermicompost increased the yield of the essential oil obtained from *D. moldavica* [21,23,26], but the use of vermicompost also resulted in an increase in soil microbial activity [26], indicating that not only yield and yield quality, but also soil sustainability, are positively affected by these organic fertilizers. The results of the present study provide additional support to these facts, but also indicate that special attention should be paid to the characteristics of substrates used for the preparation of vermicompost-containing mixes.

A direct positive effect of vermicompost for plants has been associated with the increased supply of plant-available mineral nutrients as well as the presence of plant growth-stimulating substances [5]. While it is reasonable to suggest that the addition of vermicompost to soil or any other substrate used for plant cultivation results in physico-chemical or biological interactions between various components, it is not clear if these interactions could affect plants, as no studies so far have addressed the question of the influence of substrate type on the effect of vermicompost on plants. When in a previous study *D. moldavica* plants were cultivated in a relatively nutrient-poor soil, the plant shoot biomass linearly increased with an increasing vermicompost amendment rate up to 40%, mostly because of the additional supply of mineral nutrients [27]. In the present study, substrates with a relatively high content of plant-available mineral nutrients were used, with near-optimum concentrations of elements for peat substrate and only some shortage of N for commercial garden soil (Table 1). As a result, the growth of *D. moldavica* plants was relatively similar in pure substrates, with a 16% decrease of fresh mass of stems in the case of garden soil (Figure 4B) that was obviously only related to a lower tissue water content (Figure 6B). Moreover, the additional application of mineral fertilizer three times during the cultivation period increased plant biomass only by 25% in peat and 15% in soil (Figures 4 and 5), indicating that the concentration of plant-available mineral nutrients in both substrates was close to optimum, initially, at least. Biomass increase by vermicompost amendment was much higher, at 114% and 98%, in comparison to control in the case of a 20% amendment rate, for peat- and soil-grown plants, respectively.

However, for a 30% amendment rate, these values were 148% and 68% for peat- and soil-grown plants, respectively. Consequently, the addition of an identical amount of vermicompost resulted in a poorer growth response of plants in commercial garden soil as a substrate in comparison to peat, but an increase of the amendment rate from 20% to 30% resulted in some growth inhibition for these plants.

Poor crop growth in substrates at high rates of amendment with organic fertilizers has been associated with a too high amount of soluble salts, possibly leading to osmotic or ionic stress [39,40]. Root exudates of some plant species (i.e., *Thymus vulgaris*) can facilitate the mineralization of organic matter, further enhancing the content of soluble salts in organic fertilizer-amended soil [27,41]. However, the growth suppression of *D. moldavica* plants in garden soil, amended with 30% vermicompost, could not be due to excessive soluble salt concentration, as the initial EC in cultivation substrate in the case of the vermicompost-amended peat substrate was even higher than that for the amended soil (Figure 1). In addition, no metabolic toxicity could be proposed, as plants grown in soil amended with 30% vermicompost had values of F_v/F_m, which is an indicator of plant stress, around 0.82, corresponding to a highly optimal physiological state of photochemistry [42]. It seems that some other characteristics, apart from the soluble salt content, in the two substrate systems could account for the differences in the growth responses of *D. moldavica* plants in those substrates amended with an equal amount of vermicompost. One possible target characteristic could be related to microbial activity, which is evidently significantly higher in commercial garden soil in comparison to that in peat substrate. Vermicompost itself is a very rich source of bacterial and fungal diversity, and the total number of cultivable filamentous fungi shows a significant impact on the growth-stimulating activity of vermicompost extracts [43]. It seems that some type of unidentified active factor from commercial garden soil changed some characteristic of vermicompost important for plant growth stimulation, through an inactivation of hormone-like substances or production of growth inhibitors.

The stimulation of water accumulation in plant shoots is one of the main effects of the application of humic substances for plants, besides stimulation of root growth [44], and this effect has been also found for vermicompost application [27]. In the present study, the water content of leaves significantly increased only at a 30% vermicompost amendment rate for both substrates (Figure 6A); however, for the stems there was a significant effect only in the case of a 30% substitution in peat substrate (Figure 6B). Most likely, a higher degree of water accumulation in plant tissues reflects an increase in mineral nutrient availability and is related to the efficient decrease of ionic strength and/or the osmotic potential in cells [45]. Similarly, the increased tissue water content of halophytic species under high soil salinity is probably associated with extensive vacuolar development because of ion accumulation [46].

The chlorophyll *a* fluorescence measurements provide complex information on the physiological status of PSII, both reaction centers and antenna, and the components of electron transfer chain at both donor and acceptor sides, as well as the effect of light-independent reactions [30]. Among them, PI is an essence of photochemical reactions, incorporating information on energy fluxes at the most crucial steps of energy transfer during the photochemical reactions of photosynthesis. It needs to be stressed that the absolute values of PI measurement cannot be used for the characterization of any sample, but only the changes in these parameters in the sample adds meaning to these measurements [47].

In respect to the usefulness of physiological indices for monitoring the status of plants and predicting plant yield, it needs to be stressed that changes in chlorophyll concentration reflect important metabolic changes due to changed conditions, rather than acting as a cause of these changes. In contrast, the variation of activity in the photochemical processes of photosynthesis because of changed conditions can directly lead to a decrease in photosynthesis rate and a reduced plant growth and yield. Chlorophyll concentration increased in field-grown *D. moldavica* plants due to compost application, but there was no relationship with the compost dose applied [21]. Similar results were obtained in another field study with *D. moldavica* [23]. In the present experiments, the chlorophyll concentration was positively affected by a vermicompost amendment in a concentration-dependent manner, but this effect appeared relatively late during a cultivation period, when the general decreasing trend of changes in chlorophyll content was evident (Figure 8). Therefore, it is most likely that the increased chlorophyll concentration in the leaves of *D. moldavica* plants grown in vermicompost-amended substrate reflected delayed senescence of these plants, due to a prolonged supply of mineral nutrients.

In the present study, large differences were found between three groups of fluorescence-derived parameters: (i) F_v/F_m and F_v/F_0, (ii) PI_{inst} and PI_{abs}, and (iii) PI_{total} (Figures 9–13). In addition, these groups showed a variable level of predictability, with respect to differences in plant yield due to a pronounced variation, in correlation with time (Figure 14). While both PI_{abs} and PI_{total} are calculated from temporal changes in fluorescence emission on an absorption basis, PI_{total} also involves changes in energy fluxes as related to the reduction of PSI end electron acceptors [47]. Thus, the extreme decrease of PI_{total} with plant age seems to be related to the diminished efficiency of electron transfer at the PSI side. Similarly, PI_{total} has been shown to represent a sensitive indicator of leaf senescence, especially in comparison to PI_{abs} [48]. As F_v/F_0 values during the second half of the vegetation period had a good predictability of yield of *D. moldavica*, it seems that mostly photochemical reactions at the donor side of PSII, including the activity of the water-splitting complex, were important as yield determinants during that particular period of plant development.

In conclusion, the incorporation of vermicompost for the cultivation of *D. moldavica*, even in substrate mixes with a relatively high and balanced composition of plant-available nutrients, benefits plant growth, physiological status, and biomass yield. It appears that organic fertilizers and especially earthworm-produced vermicomposts are useful substrate components for the sustainable cultivation of *D. moldavica*. The nondestructive chlorophyll fluorescence analysis can be successfully used to predict biomass accumulation of *D. moldavica* plants grown in different substrates. It appears that, to gain maximal benefits from vermicompost application in a form of increased plant yield and quality, it is necessary to explore interactions between vermicompost and other substrates leading to possible changes in the quality-related characteristics of vermicompost in substrate mixes. Most importantly, a detailed analysis of microbial processes and their effect on the quality characteristics of vermicompost needs to be investigated.

Author Contributions: A.O. and G.I. conceived and designed the study; A.O., U.A.-O. and G.I. performed experiments and gathered data; A.O. and G.I. analyzed and interpreted the data; A.O. and G.I. wrote the manuscript. All authors have approved the manuscript for publication. G.I. takes responsibility for the integrity of the work as a whole. All authors have read and agreed to the published version of the manuscript.

Funding: This research received no external funding.

Institutional Review Board Statement: Not applicable.

Informed Consent Statement: Not applicable.

Data Availability Statement: All data reported here is available from the authors upon request.

Acknowledgments: We thank Jānis Freibergs (Eko Zeme, Latvia) for providing the vermicompost sample used in the present study.

Conflicts of Interest: The authors declare no conflict of interest.

References

1. Cakmak, I. Plant nutrition research: Priorities to meet human needs for food in sustainable ways. *Plant Soil* **2002**, *247*, 3–24. [CrossRef]
2. Lazcano, C.; Gómez-Brandón, M.; Domínguez, J. Comparison of the effectiveness of composting and vermicomposting for the biological stabilization of cattle manure. *Chemosphere* **2008**, *72*, 1013–1019. [CrossRef]

3. Gopal, M.; Gupta, A.; Sunil, E.; Thomas, G.V. Amplification of plant beneficial microbial communities during conversion of coconut leaf substrate to vermicompost by *Eudrilus* sp. *Curr. Microbiol.* **2009**, *59*, 15–20. [CrossRef] [PubMed]
4. Kaur, A.; Singh, J.; Vig, A.P.; Dhaliwal, S.S.; Rup, P.J. Cocomposting with and without *Eisenia fetida* for conversion of toxic paper mill sludge to a soil conditioner. *Bioresour. Technol.* **2010**, *101*, 8192–8198. [CrossRef] [PubMed]
5. Ievinsh, G. Review on physiological effects of vermicomposts on plants. In *Biology of Compost*; Soil Biology Volume 58; Meghvansi, M.K., Varma, A., Eds.; Springer Nature: Cham, Switzerland, 2020; pp. 63–86.
6. Aslam, Z.; Bashir, S.; Hassan, W.; Bellitürk, K.; Ahmad, N.; Niazi, N.K.; Khan, A.; Khan, M.I.; Chen, Z.; Maitah, M. Unveiling the efficiency of vermicompost derived from different biowastes on wheat (*Triticum aestivum* L.) plant growth and soil health. *Agronomy* **2019**, *9*, 791. [CrossRef]
7. Gupta, C.; Prakash, D.; Gupta, S.; Nazareno, M.A. Role of vermicomposting in agricultural waste management. In *Sustainable Green Technologies for Environmental Management*; Shah, S., Venkatramanan, V., Prasad, R., Eds.; Springer Nature: Singapore, 2019; pp. 283–295.
8. Arancon, N.Q.; Edwards, C.A.; Atiyeh, R.; Metzger, J.D. Effects of vermicompost produced from food waste on the growth and yields of greenhouse peppers. *Bioresour. Technol.* **2004**, *93*, 139–144. [CrossRef] [PubMed]
9. Pączka, G.; Mazur- Pączka, A.; Garczyńska, M.; Hajduk, E.; Kostecka, J.; Bartkowska, I.; Butt, K.R. Use of vermicompost from sugar beet pulp in cultivation of peas (*Pisum sativum* L.). *Agriculture* **2021**, *11*, 919. [CrossRef]
10. Schröder, C.; Häfner, F.; Larsen, O.C.; Krause, A. Urban organic waste for urban farming: Growing lettuce using vermicompost and thermophilic compost. *Agronomy* **2021**, *11*, 1175. [CrossRef]
11. Karlsons, A.; Osvalde, A.; Andersone-Ozola, U.; Ievinsh, G. Vermicompost from municipal sewage sludge affects growth and mineral nutrition of winter rye (*Secale cereale*) plants. *J. Plant Nutr.* **2016**, *39*, 765–780. [CrossRef]
12. Hussain, N.; Abbasi, S.A. Efficacy of the vermicomposts of different organic wastes as "clean" fertilizers: State-of-the-art. *Sustainability* **2018**, *10*, 1205. [CrossRef]
13. Blouin, M.; Barrere, J.; Meyer, N.; Lartigue, S.; Barot, S.; Mathieu, J. Vermicompost significantly affects plant growth. A meta-analysis. *Agron. Sustain. Dev.* **2019**, *39*, 34. [CrossRef]
14. Pierre-Louis, R.; Kader, M.A.; Desai, N.M.; John, E.H. Potentiality of vermicomposting in the South Pacific island countries: A review. *Agriculture* **2021**, *11*, 876. [CrossRef]
15. Vuković, A.; Velki, M.; Ečimović, S.; Vuković, R.; Čamagajevac, I.Š.; Lončarić, Z. Vermicomposting–facts, benefits and knowledge gaps. *Agronomy* **2021**, *11*, 1952. [CrossRef]
16. Soni, R.; Sharma, A. Vermiculture technology: A novel approach in organic farming. *Indian Hortic. J.* **2016**, *6*, 150–154.
17. Lim, S.L.; Wu, T.Y.; Lim, P.N.; Shak, K.P.Y. The use of vermicompost in organic farming: Overview, effects on soil and economics. *J. Sci. Food Agric.* **2015**, *95*, 1143–1156. [CrossRef]
18. Warman, P.R.; AngLopez, M.J. Vermicompost derived from different feedstocks as a plant growth medium. *Bioresour. Technol.* **2010**, *101*, 4479–4483. [CrossRef] [PubMed]
19. Greco, C.; Comparetti, A.; Fascella, G.; Febo, P.; La Placa, G.; Saiano, F.; Mammano, M.M.; Orlando, S.; Laudicina, V.A. Effects of vermicompost, compost and digestate as commercial alternative peat-based substrates on qualitative parameters of *Salvia officinalis*. *Agronomy* **2021**, *11*, 98. [CrossRef]
20. Aćimović, M.; Sikora, V.; Brdar-Jokanović, M.; Kiprovski, B.; Popović, V.; Koren, A.; Puvača, N. *Dracocephalum moldavica*: Cultivation, chemical composition and biological activity. *J. Agron. Technol. Eng. Manag.* **2019**, *2*, 152–167.
21. Hussein, M.S.; El-Sherbeny, S.E.; Khalil, M.Y.; Nagib, N.Y.; Aly, S.M. Growth characters and chemical constituents of *Dracocephalum moldavica* L. plants in relation to compost fertilizer and planting distance. *Sci. Hortic.* **2006**, *1108*, 322–331. [CrossRef]
22. Mostafa, H.S. Complementary effect between compost rate and ascorbic acid concentration on enhancing dragonhead (*Dracocephalum moldavica*) plant on growth and productivity. *Middle East J. Agric.* **2018**, *7*, 1811–1818.
23. Badawy, E.S.M.; El-Sherbeny, S.; Hussein, M.S.; Swaefy, H.M.F.; Amer, H.M. Effect of organic manure on growth, production and active ingredients in dragonhead (*Dracocephalum moldavica* L.). *Medic. Aromat. Plant Sci. Biotechnol.* **2009**, *3*, 9–13.
24. Nejatzadeh-Barandozi, F.; Shahvaladi, E.H.; Gholami-Borujeni, F. Nitrogen fertilization and microelements influences growth index and yield in *Dracocephalum moldavica* L. *Bulg. J. Agric. Sci.* **2015**, *21*, 266–269.
25. Nasiri, Y. Crop productivity and chemical compositions of dragonhead (*Dracocephalum moldavica* L.) essential oil under different cropping patterns and fertilization. *Industr. Crops Prod.* **2021**, *171*, 113920. [CrossRef]
26. Rezaei-Chiyaneh, E.; Amirnia, R.; Chiyaneh, S.F.; Maggi, F.; Barin, M.; Razavi, B.S. Improvement of dragonhead (*Dracocephalum moldavica* L.) yield quality through a coupled intercropping system and vermicompost application along with maintenance of soil microbial activity. *Land Degrad. Dev.* **2021**, *32*, 2833–2848. [CrossRef]
27. Ievinsh, G.; Andersone-Ozola, U.; Zeipiņa, S. Comparison of the effects of compost and vermicompost soil amendments in organic production of four herb species. *Biol. Agric. Hortic.* **2020**, *36*, 267–282. [CrossRef]
28. Andersone, U.; Druva-Lūsīte, I.; Ieviņa, B.; Karlsons, A.; Ņečajeva, J.; Samsone, I.; Ievinsh, G. The use of nondestructive methods to assess a physiological status and conservation perspectives of *Eryngium maritimum* L. *J. Coast. Conserv.* **2011**, *15*, 509–522. [CrossRef]
29. Matisons, R.; Krišāns, O.; Jansons, Ā.; Kondratovičs, T.; Elferts, D.; Ievinsh, G. Norway spruce seedlings from an Eastern Baltic provenance show tolerance to simulated drought. *Forests* **2021**, *12*, 82. [CrossRef]

30. Kalaji, H.M.; Jajoo, A.; Oukarroum, A.; Brestic, M.; Zivcak, M.; Samborska, I.A.; Cetner, M.D.; Łukasik, I.; Goltsev, V.; Ladle, R.J. Chlorophyll *a* fluorescence as a tool to monitor physiological status of plants under abiotic stress conditions. *Acta Physiol. Plant.* **2016**, *38*, 102. [CrossRef]
31. Strand, M. Effect of mineral nutrient content on oxygen exchange and chlorophyll *a* fluorescence in needles of Norway spruce. *Tree Physiol.* **1997**, *17*, 221–230. [CrossRef]
32. Kastori, R.; Plesnicar, M.; Arsenijevic-Maksimovic, I.; Petrovic, N.; Pankovic, D.; Sakac, Z. Photosynthesis, chlorophyll fluorescence, and water relations in young sugar beet plants as affected by sulfur supply. *J. Plant Nutr.* **2000**, *23*, 1037–1049. [CrossRef]
33. Wang, H.; Liu, R.L.; Jin, J.Y. Effects of zinc and soil moisture on photosynthetic rate and chlorophyll fluorescence parameters of maize. *Biol. Plant.* **2009**, *53*, 191–194. [CrossRef]
34. Horaczek, T.; Dąbrowski, P.; Kalaji, H.M.; Baczewska-Dąbrowska, A.H.; Pietkiewicz, S.; Stępień, W.; Gozdowski, D. JIP-test as a tool for early detection of the macronutrients deficiency in *Miscanthus* plants. *Phtosynthetica* **2020**, *58*, 507–517. [CrossRef]
35. Osvalde, A. Optimization of plant mineral nutrition revisited: The roles of plant requirements, nutrient interactions, and soil properties in fertilization management. *Environ. Exp. Biol.* **2011**, *9*, 1–8.
36. Strasser, R.J.; Srivastava, A.; Tsimilli-Michael, M. The fluorescence transient as a tool to characterise and screen photosynthetic samples. In *Probing Photosynthesis: Mechanisms, Regulation and Adaptation*; Yunus, M., Pathre, U., Mohanty, P., Eds.; Taylor & Francis: London, UK, 2000; pp. 445–483.
37. Banks, J.M. Continuous excitation chlorophyll fluorescence parameters: A review for practicioners. *Tree Physiol.* **2017**, *37*, 1128–1136. [CrossRef]
38. McDonald, J.H. *Handbook of Biological Statistics*, 3rd ed.; Sparky House Publishing: Baltimore, MD, USA, 2014; p. 299.
39. Tiquia, S.M. Reduction of compost phytotoxicity during the process of decomposition. *Chemosphere* **2010**, *79*, 506–512. [CrossRef] [PubMed]
40. Jakubus, M. A comparative study of composts prepared from various organic wastes based on biological and chemical parameters. *Agronomy* **2020**, *10*, 869. [CrossRef]
41. Kramer, C.; Gleixner, G. Variable use of plant- and soil-derived carbon by microorganisms in agricultural soils. *Soil Biol. Biochem.* **2006**, *38*, 3267–3278. [CrossRef]
42. Björkman, O.; Demmig, B. Photon yield of O_2 evolution and chlorophyll fluorescence characteristics at 77 K among vascular plants of diverse origins. *Planta* **1987**, *170*, 489–504. [CrossRef] [PubMed]
43. Grantina-Ievina, L.; Andersone, U.; Berkolde-Pīre, D.; Nikolajeva, V.; Ievinsh, G. Critical tests for determination of microbiological quality and biological activity in commercial vermicompost samples of different origins. *Appl. Microbiol. Biotechnol.* **2013**, *97*, 10541–10554. [CrossRef]
44. Piccolo, A.; Celano, G.; Pietramellara, G. Effects of fractions of coal-derived humic substances on seed germination and growth of seedlings (*Lactuga sativa* and *Lycopersicon esculentum*). *Biol. Fertil. Soils* **1993**, *16*, 11–15. [CrossRef]
45. Ievinsh, G.; Andersone-Ozola, U.; Sieriņa, A. Development and physiological performance of hydroponically-grown ornamental indoor plants in respect to their potential use in botanical biofilters: Effect of mineral nutrient availability. *Proc. Latvian Acad. Sci. B* **2021**, in press.
46. Ogburn, R.M.; Edwards, E.J. The ecological water-use strategies of succulent plants. In *Advances in Botanical Research*; Kader, J.-C., Delseny, M., Eds.; Academic Press: Burlington, MA, USA, 2010; Volume 55, pp. 179–225.
47. Tsimilli-Michael, M. Revisiting JIP-test: An educative review on concepts, assumptions, approcimations, definitions and terminology. *Photosynthetica* **2020**, *58*, 275–292. [CrossRef]
48. Vuletić, M.V.; Španić, V. Characterization of photosynthetic performance during natural leaf senescence in winter wheat: Multivariate analysis as a tool for phenotypic characterization. *Photosynthetica* **2020**, *58*, 301–313. [CrossRef]

agriculture

Article

A Comparative Study of the Fertilizer-Cum-Pesticide Effect of Vermicomposts Derived from Cowdung and the Toxic Weed Lantana

Naseer Hussain [1], Tabassum-Abbasi [2], Tasneem Abbasi [1] and Shahid Abbas Abbasi [1,*]

[1] Centre for Pollution Control & Environmental Engineering, Pondicherry University, Chinakalapet, Puducherry 605014, India; naseer890@gmail.com (N.H.); t.abbasi.cpe@pondiuni.edu.in (T.A.)

[2] School of Engineering, University of Petroleum and Energy Studies, Bidholi, Dehradun 248007, India; tabbasi@ddn.upes.ac.in

* Correspondence: prof.abbasi.cpe@pondiuni.edu.in

Abstract: The effect of vermicomposts, derived either from cowdung or the pernicious invasive plant lantana (*Lantana camara*), has been assessed on the seed germination, plant growth, fruit yield, quality of the produce, and disease resistance of a common vegetable, ladies finger (*Abelmoschus esculentus*).Seeds of *A. esculentus* were germinated and grown in soil fertilized with 0, 2.5, 3.75 and 5 t ha^{-1} of lantana or cowdung vermicompost for 4 months. It was seen that the lantana vermicompost performed at par or better than the cowdung vermicompost in terms of most of the growth and yield parameters observed. Both the vermicomposts encouraged the germination, growth as well as the yield of ladies fingers. The fruits harvested from the vermicompost-treated plots had greater concentrations of minerals, proteins and carbohydrates than the control plants. Vermicomposts also reduced the incidence of pest attacks on the plants. The results confirm that vermicomposting destroys the harmful ingredients of lantana and turns it into as good a biofertilizer, perhaps even better than the vermicompost of cow-dung. The very large quantities of lantana biomass that is generated in the tropical and sub-tropical regions of the world every year, which presently go to waste, now appear capable of becoming a source of organic fertilizer.

Keywords: allelopathy; lantadene; weeds control; vermicomposting; organic fertilizer

1. Introduction

Lantana (*L. camara*), which is acknowledged as one among the 100 most invasive and colonizing of the world's weeds [1], has become a major threat to agriculture and forest ecosystems [2,3].It has the ability to grow in widely varying environmental conditions [4,5], often forming large, impenetrable, thickets.Due to its colonizing ability lantana monopolizes the use of light, water, and nutrients in the areas it invades, at the expenses of multi-species vegetation, causing great harm to biodiversity [6]. Being rich in toxic chemicals such astriterpene acids, lantadene A (rehmannic acid) and lantadene B, lantana induces cholestasis, hepatotoxicity and mortality in animals who graze on its foliage [7–9]. Lantana is also strongly allelopathic and restricts the growth of surrounding vegetation [10]. Even though efforts have been made since several decades to control lantana by mechanical, chemical, biological or hybrid means [11], no enduring success has been achieved till now and lantana continues to overrun ever new territories. Attempts to use lantana as a feedstock for producing cellulose, ethanol, drugs, or compost could engage only a small fraction of its biomass [12–14] with no market penetration so far. Therefore, it is imperative that an economically viable product of large global demand is developed using lantana.

In nature earthworms feed voraciously on the debris of all species of plants, including those known to be toxic to vertebrates.It is believed that these animals carry a class of rare surface-active metabolites in their bodies, which have been termed 'drilodefensins' [15]. These compounds cancelledthe inhibitory effects of polyphenols and other toxic chemicals present in plants like lantana on earthworm gut enzymes and enable the earthworms to tolerate high levels of polyphenols if present in their diet.As a result, the earthworms are able to feed on a large variety of phytomass, including streams with high levels of polyphenols.

We have earlier reported [16,17] that even though in nature epigeic and anecic earthworms principally feed upon plant debris—and much less animal dropping in proportion—controlled vermicomposting on large scale has so far been limited to animal manure. We have explained the reasons and have described the concept of high-rate vermicomposting developed by us along with the technological interventions done by us which has made it possible to vermicompost lantana and other weeds on a large scale [16,18].

We have successfully used the epigeic earthworm *Eisenia foetida* for vermicomposting lantana [17]. Extensive investigations to characterize the lantana vermicompost (LVC) using Fourier transform infrared spectroscopy, thermal gravimetry, differential calorimetric analysis, gas chromatography, and scanning electron micrography (SEM) have revealed intense mineralization of the organic matter, degradation of lignocellulosic materials and polyphenols, reduction of toxic and allelopathic compounds (phenols and sesquiterpene lactones) in the course of lantana's vermicomposting. SEM has reflected strong disaggregation of the organic matter content in LVC compared to the lantana matrices. Further, in a controlled study, Hussain et al. [19], have observed that LVC enhanced the germination of the seeds, and early growth of the seedlings of ladies finger, green gram (*Vigna radiata*) and cucumber (*Cucumis sativus*) when used at appropriate concentrations in soil. However, beyond certain level lantana vermicompost had shown adverse effects. This had raised apprehensions as to whether LVC behaves differently from cow-dung vermicompost (CDVC). It was, therefore, decided to compare the effects of LVC and CDVC under identical conditions. Accordingly, we have carried out this study in which the effect of CDVC has been compared with that of lantana vermicompost on the growth, fruition and quality of the ladies finger produce, in a field-scale study.

2. Materials and Methods

2.1. Soil and the Vermicomposts

The studies were conducted in a field situated within the boundary of Pondicherry University, India. The study area lies on the eastern coast of the peninsular South India, at 11°56′ N, and 79°53′ E. The studies were performed in the months spanning February–May which is the season known to be the most suited for the cultivation of ladies finger in the place where the authors work. The soil used in the study was obtained from within the Pondicherry University campus; its characteristics are presented in Table 1. Cowdung was procured from the local farmers in the vicinity of Pondicherry University campus and lantana was harvested from its stands in and around the campus. Vermicomposts from both were generated using the concept of high-rate vermicomposting and the FLUVTS machine as elaborated earlier [20] for obtaining vermicompost from paper waste. In both cases the earthworm dropping, which are easily distinguishable and separable from the parent substrate, were identified as the vermicompost.

Table 1. Composition of the soil and the vermicomposts deployed in the present study.

Aspect	Lantana VC	Cow Dung VC	Soil
Total nitrogen, g/Kg	19.6 ± 2	23 ± 2.7	0.69 ± 0.05
Available phosphorus, g/Kg	7.5 ± 0.8	5.3 ± 0.4	0.26 ± 0.04
Available Potassium, g/Kg	18.5 ± 1.8	14.8 ± 2.1	0.81 ± 0.08
Total organic carbon, g/Kg	283 ± 18	258 ± 26	8.79 ± 0.63
C/N	14.4 ± 0.6	11.2 ± 0.7	13.9 ± 0.9
Particle density, g/cm^3	1.4 ± 0.3	1.5 ± 0.1	2.7 ± 0.04
Bulk density, g/cm^3	0.35 ± 0.02	0.4 ± 0.01	1.4 ± 0.03
Water holding capacity, %	252 ± 17	235 ± 22	35 ± 3
Porosity, %	72 ± 1.9	66 ± 1.9	49 ± 1
Electrical conductivity, mmhos cm^{-1}	10.1 ± 0.24	11.7 ± 0.9	0.16 ± 0.02
pH	6.4 ± 0.17	7.2 ± 0.2	6.35 ± 0.15

2.2. Design of Experiments

Plants were grown outdoors in 50-litre LDPE (low-density polyethylene) containers filled with soil. The design of experiments consisted of the use of controls without any amendment and of vermicomposts at three levels: 2.5, 3.75, and 5 t ha^{-1} [21]. In each of the selected treatment, a total number of 175 seeds were sown in 35 bags. The *Kulemagali vendai* variety of ladies finger, which is locally available, was used. The number of seeds germinated over an 8-day period were counted to obtain germination success in terms of percentage of seeds germinated. On day nine, four seedlings from each bag were removed so as to keep a single healthy plant in each bag. Plants were allowed to grow up to 100 days. Throughout this period the bags were periodically irrigated with tap water.

2.3. Sampling and Analysis

After 100 days, the plant samples consisting of five randomly-selected whole ladies finger plants per set, were harvested for the assessment of morphological growth which was recorded in terms of mean plant height, number of leaves and branches, stem diameter, and above-ground biomass. The plant's roots were rinsed with water to clear off the adhering soils, before further analysis. Dry weight of the plants was determined by oven drying their known quantities at 105 °C to constant weights. The yield of ladies finger on the basis of pods per plant, and the length (cm), diameter (mm), and weight (g) of the pods per plant was recorded on alternate days. The chlorophyll content of the vegetable's leaves was determined on the basis of the procedure detailed by Moran and Porath [22] and Wellburn [23]. The vegetable's pods were analyzed for their content of protein, carbohydrate and ash by the Kjeldahl, Anthrone and dry ashing methods, respectively (Nielson, 2010). The total solids of the pods were determined by heating their measured quantities at 105 °C to constant weights, as per the procedure of Nielson [24].

To measure the pH and the electrical conductivity (EC) of the vermicast and the soil, their 1:2 (*w*/*v*) suspensions were prepared in water using Digison™ digital pH meter 7007 and ET™611E EC meter, respectively. The bulk density, particle density, and total porosity of the soil and vermicast samples were measured following the procedure reported by Carter and Gregorich [25]. The two matrice's capacity to hold water was determined by measuring their gravimetric water content following and saturation of samples and draining of the excess water [26].

Total organic carbon was estimated using the modified dichromate redox method for respective weeds and their vermicast as described by Heanes [27]. Determination of total nitrogen was done with the modified Kjeldahl method [28] employing a KelPlus™

instrument. Extractable/available potassium and phosphorus were determined employing Elico^{TM} CL378 flame photometer and ammonium molybdate-ascorbic acid method, respectively, after the samples were extracted with Mehlich 3 solution [25].

During the experiments some of the vegetable plants were found to have been infested with leaf miners and leaf spot diseases. These infestations were caused by plant pests *Liriomyza* spp. and fungus *Alternaria alternate*, respectively [29,30]. In case of severity, the leaf spot disease generates concentric dark brown spots on the leaves, eventually causing the death of the leaves [31]. Some of the ladies finger pods were found to be infested with fruit borer *Eariasvittella*. The extent of infestation was calculated as percentage of the weight of the effected fruits with reference to the total fruit weight in each treatment.

2.4. Assessment of Levels of Significance

The effect of the vermicompost treatments was compared with the controls using statistical test of one-way analysis of variance (ANOVA). The overall effect of LVC and CDVC on all the morphological and biochemical aspects of ladies finger was compared by a two-way ANOVA. Comparisons were made as types of vermicomposts (VC), concentration of vermicomposts (N) and their interactions.

3. Results and Discussion

3.1. Seed Germination

The findings are summarized in Figure 1. Vermicompost treatments significantly enhanced the seed germination compared with the controls (Figure 1a), however no statistically significant variation was seen between the effects of the cowdung and the lantana vermicompost treatments. The highest germination success (95%) was seen in 5 t ha^{-1} lantana vermicompost (LVC) treatment. The next best success (94%) occurred in the 3.75 t ha^{-1} cowdung vermicompost (CDVC) treatment. Even though seed germination is primarily an internally regulated mechanism which is governed by the genotype of the plant, several environmental factors and fertilization regimes can also alter the germination success [18]. Several of the studies have suggested that besides the plant hormones and phenolic compounds, increased nitrate and ammonium concentrations in the vermicompost play a strong role in seed germination [32,33].

3.2. Plant Growth

Ladies finger plants grown in VC amended soils have shown enhanced growth in terms of all the variables recorded (Figure 1b–g). Within the range of vermicompost concentrations explored by us, the trend of positive effect was: greater the vermicompost application more the benefit. Apart from the number of leaves in CDVC, all trends had the pattern 5 t > 3.75 t > 2.5 t ha^{-1} > control. Except for the length of the roots, the growth of ladies finger went up profusely even when the concentration of both the vermicomposts was increased only marginally (from zero to 2.5 t ha^{-1}). Similar observations were recorded for flowering, where higher LVC treatments yielded a greater number of flowers and induced earlier flowering relative to the controls and the lower LVC treatments. In case of CDVC, the 3.75 t ha^{-1} treatment performed better than other treatments (Table 2).

In comparison to CDVC, the shoot length and the plant biomass were significantly higher in the ladies finger plant grown in LVC amended soil; however there was no statistically significant difference vis a vis shoot diameter and the number of branches. As elucidated by Hussain and Abbasi [18], vermicompost amendment in soil enhances the available nutrient content of the soil, besides making the soil porosity, density, and water holding capacity more plant-friendly. In addition, soils amended with vermicomposts were seen to be rich in fulvic and humic acids, and plant hormones [34], which apparently boost the growth of plants compared to the controls. The results of the present investigation show that in some aspects LVC has outperformed CDVC while in some other aspects no significant difference was seen between the two. This makes it evident that lantana loses its toxic and allelopathic constituents during its vermicomposting and the resultant vermicompost, has positive influence on the growth of ladies finger. Equally significant is the finding that the positive influence matches—at times even surpasses—that of CDVC.

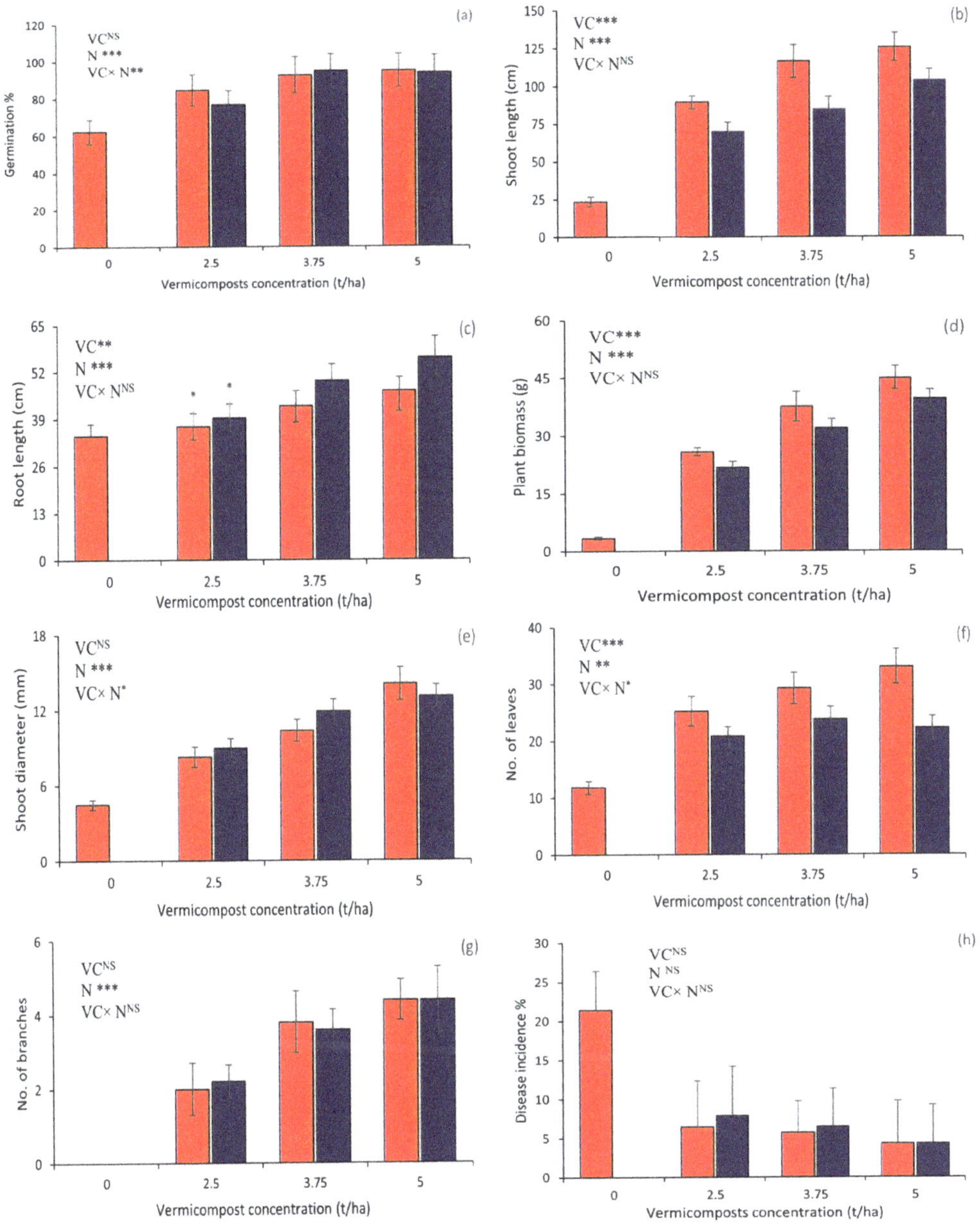

Figure 1. Effect of LVC and CDVC on ladies fingerin terms of (**a**) germination success; (**b**) length of shoots; (**c**) length of roots; (**d**) plant biomass; (**e**) shoot diameter; (**f**) number of leaves; (**g**) number of branches; and (**h**) disease incidence. All the bars carry range of standard deviation. Bars topped with an asterisk indicate that the corresponding numbers do not differ significantly from the controls at $p \leq 0.05$. N indicate the vermicompost treatments.

Table 2. Flowering and yield of *A. esculentus* plants grown in soil fertilized with different levels of lantana and cowdung vermicomposts. The numbers which do not differ significantly from controls ($p < 0.05$) carry an asterisk. Single, double, and triple stars indicate the significance levels at $p < 0.5$, <0.01 and <0.001, respectively.

<table>
<tr><th rowspan="2">Parameters Observed</th><th rowspan="2">Type of VC</th><th colspan="4">Vermicompost Concentrations (t/ha)</th><th colspan="3">ANOVA</th></tr>
<tr><th>0</th><th>2.5</th><th>3.75</th><th>5</th><th>Type of Vermicompost (VC)</th><th>Concentration of Vermicompost (N)</th><th>VC*N</th></tr>
<tr><td rowspan="2">Days to flower</td><td>LCVC</td><td rowspan="2">52.7 ± 4.85</td><td>43.2 ± 2.30</td><td>39.0 ± 3.37</td><td>37.3 ± 2.41</td><td rowspan="2">NS</td><td rowspan="2">***</td><td rowspan="2">NS</td></tr>
<tr><td>CDVC</td><td>43.3 ± 2.75</td><td>38.6 ± 2.84</td><td>39.3 ± 2.79</td></tr>
<tr><td rowspan="2">No. of flowers</td><td>LCVC</td><td rowspan="2">2.9 ± 0.32</td><td>9.0 ± 1.05</td><td>16.3 ± 1.16</td><td>18.0 ± 2.16</td><td rowspan="2">***</td><td rowspan="2">***</td><td rowspan="2">***</td></tr>
<tr><td>CDVC</td><td>8.3 ± 0.95</td><td>12.8 ± 1.32</td><td>10.1 ± 0.88</td></tr>
<tr><td rowspan="2">No. of pods</td><td>LCVC</td><td rowspan="2">1.7 ± 0.48</td><td>6.2 ± 0.63</td><td>13.7 ± 1.06</td><td>16.2 ± 2.10</td><td rowspan="2">***</td><td rowspan="2">***</td><td rowspan="2">***</td></tr>
<tr><td>CDVC</td><td>6.5 ± 0.53</td><td>10.8 ± 1.03</td><td>8.6 ± 0.52</td></tr>
<tr><td rowspan="2">Length of pods (cm)</td><td>LCVC</td><td rowspan="2">7.1 ± 0.50</td><td>10.9 ± 1.11</td><td>11.6 ± 0.94</td><td>13.1 ± 1.34</td><td rowspan="2">NS</td><td rowspan="2">**</td><td rowspan="2">*</td></tr>
<tr><td>CDVC</td><td>11.1 ± 0.98</td><td>11.7 ± 0.69</td><td>11.5 ± 1.10</td></tr>
<tr><td rowspan="2">Diameter of pods (mm)</td><td>LCVC</td><td rowspan="2">11.4 ± 0.70</td><td>15.4 ± 1.04</td><td>16.0 ± 0.96</td><td>16.3 ± 0.72</td><td rowspan="2">NS</td><td rowspan="2">*</td><td rowspan="2">NS</td></tr>
<tr><td>CDVC</td><td>15.6 ± 0.77</td><td>16.7 ± 1.00</td><td>15.9 ± 1.21</td></tr>
<tr><td rowspan="2">Weight of pods/plant (g)</td><td>LCVC</td><td rowspan="2">5.4 ± 0.50</td><td>91.9 ± 9.30</td><td>143.8 ± 8.47</td><td>170.5 ± 16.2</td><td rowspan="2">***</td><td rowspan="2">***</td><td rowspan="2">***</td></tr>
<tr><td>CDVC</td><td>61.8 ± 6.20</td><td>101.6 ± 8.98</td><td>85.7 ± 8.72</td></tr>
<tr><td rowspan="2">Yield t/ha</td><td>LCVC</td><td rowspan="2">0.5 ± 0.05</td><td>9.0 ± 0.91</td><td>14.1 ± 0.83</td><td>16.8 ± 1.60</td><td rowspan="2">***</td><td rowspan="2">***</td><td rowspan="2">***</td></tr>
<tr><td>CDVC</td><td>6.1 ± 0.61</td><td>10.0 ± 0.88</td><td>8.4 ± 0.86</td></tr>
<tr><td rowspan="2">Percentage infected fruits</td><td>LCVC</td><td rowspan="2">39.2 ± 12.39</td><td>9.3 ± 3.79</td><td>9.1 ± 5.42</td><td>8.0 ± 4.73</td><td rowspan="2">NS</td><td rowspan="2">*</td><td rowspan="2">NS</td></tr>
<tr><td>CDVC</td><td>13.4 ± 6.46</td><td>7.6 ± 2.77</td><td>7.6 ± 3.63</td></tr>
</table>

3.3. Yield and Biochemical Aspects

Vermicompost treatments are seen to have significantly enhanced the yield of the ladies finger pods as reflected in the average numbers and weights of pods per plant, and the average length and diameter of the pods (Table 2). In comparison to the CDVC, LVC had significantly higher number of pods per plant. It also led to pods of higher average weight. However, no significant difference was seen in case of length and diameter of the pods. Vermicompost treatments had also significantly increased the concentrations of chlorophyll and carotenoids in the ladies finger leaves, and the total solids and ash content of its fruits in comparison to the control plots (Figure 2a–d). No statistically significant difference, however, was seen between the LVC and the CDVC in terms of influence on chlorophyll, carotenoids, total solids, protein and carbohydrates content (Figure 2e–f). These gains, like the plant growth parameters, can perhaps be attributed to the increased plant available nutrients in soil fortified with vermicomposts, compared to the controls. This is consistent with similar effect reported when manure−based vermicomposts were deployed [35,36]. Overall, LVC appears to be as beneficial for the cultivation of ladies finger as CDVC.

3.4. Disease Incidence

Both the vermicomposts were able to induce disease resistance in the test plants (Figure 1h, Table 2). In terms of reducing the incidence of disease, LVC has performed marginally better than CDVC; however, the difference was not statistically significant. The fractions of infected fruits was lesser in CDVC treatments of 3.75 and 5 t ha^{-1} than in the corresponding LVC applications. However, again, the difference was not statistically significant. In a recently published review, Hussain and Abbasi [18] have documented a number of scientific studies reporting the positive role of manure-based vermicomposts in reducing pests and disease in several botanical species. The present work shows that LVC also possesses a similar virtue.

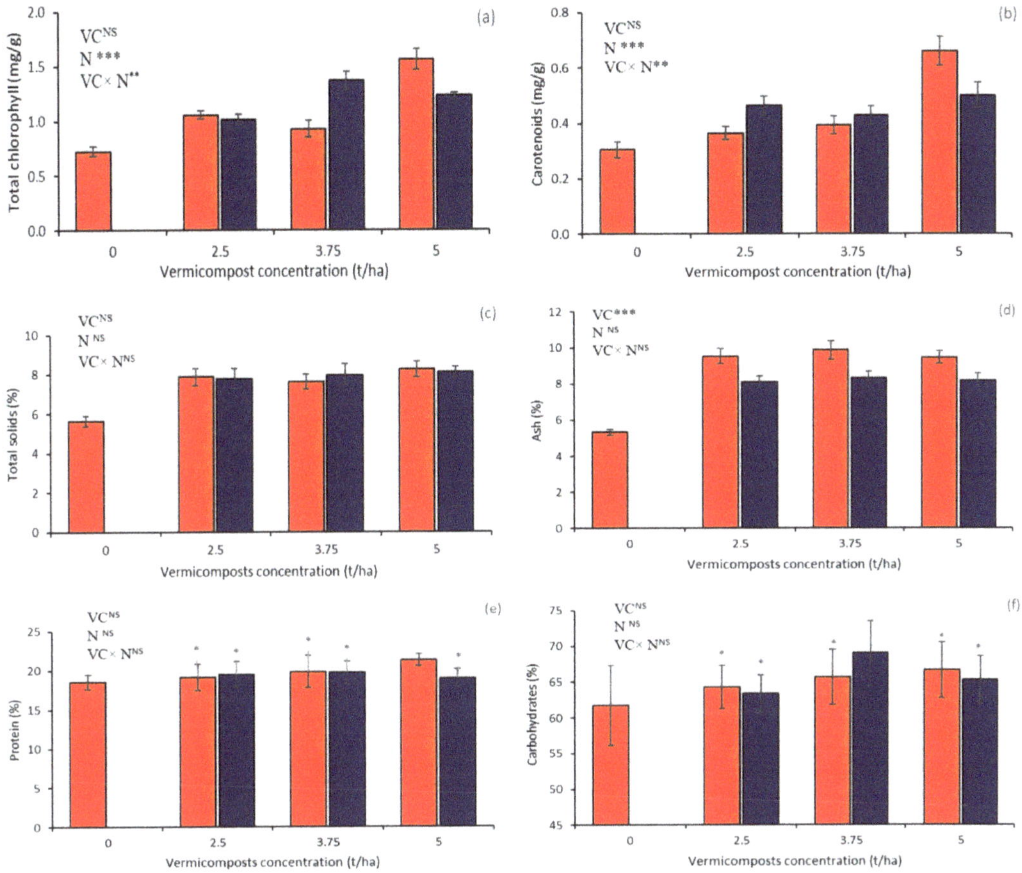

Figure 2. Effect of LVC and CDVC on ladies fingerin terms of (**a**) total chlorophyll in the leaves; (**b**) carotenoids in the leaves; (**c**) total solids; (**d**) ash; (**e**) protein in pods; and (**f**) carbohydrates in pods. All the bars carry range of standard deviation. Bars topped with an asterisk indicate that the corresponding numbers do not differ significantly from the controls at $p \leq 0.05$. N indicate the vermicompost treatments.

Previous reports on pathogen-protecting attribute of manure–based vermicomposts reveal that better nutrient availability, and presence of antimicrobial compounds such as flavonoids, phenols and humic acids in the vermicomposts, are the likely factors that may have imbibed the vermicomposts with the ability to resist pathogens [37]. Evidently these beneficial attributes are also present in LVC.

4. Summary and Conclusions

A comparative study on the effects of vermicomposts derived from lantana (LVC) and cowdung (CDVC) was carried out in terms of success in seed germination, seedling growth, yield of fruits, fruit quality and plant pathology of ladies finger (*Abelmoschusesculentus*). Contrary to the apprehensions that lantana being a toxic and allelopathic weed, its vermicompost may be unfriendly to other species of plants and the soil, LVC manifested no such negative attribute. Rather, LVC, like CDVC, enhanced the fraction of seeds that germinated, promoted the growth of the ladies finger plants, increased the fruit yield, improved the chlorophyll and carotenoid levels, and induced resistance against pests and disease, in comparison to the controls. In most of the aspects LVC had an equally beneficial, if not better, effect than CDVC. The findings add credence to the possibility that the lantana phytomass—of which enormous quantities are generated every year in the tropical and sub-tropical world—can serve as feedstock for producing much-in-demand organic fertilizer in the form of LVC.

Author Contributions: Conceptualization, N.H. and S.A.A.; Methodology, N.H. and S.A.A.; software, T.-A.; validation, T.A. and T.-A.; formal analysis, N.H. and T.-A.; investigation, N.H., T.-A. and T.A.; resources, S.A.A.; data curation, N.H. and S.A.A.; writing—original draft preparation, T.-A. and T.A.; writing—review and editing, N.H. and S.A.A.; visualization, T.-A.; supervision, S.A.A.; project ad-ministration, S.A.A. and T.A.; funding acquisition, S.A.A. All authors have read and agreed to the published version of the manuscript.

Funding: This research received no external funding.

Institutional Review Board Statement: Not applicable.

Informed Consent Statement: Not applicable.

Conflicts of Interest: The authors declare no conflict of interest.

References

1. Agaldo, J.A. Ecology of *Lantana camara*: An invasive plant species in Amurum Forest Reserve, Jos Plateau, Central Nigeria. *Afr. J. Ecol.* **2020**, *58*, 291–298. [CrossRef]
2. Qin, L.; Li, W.C.; Liu, L.; Zhu, J.Q.; Li, X.; Li, B.Z.; Yuan, Y.J. Inhibition of lignin-derived phenolic compounds to cellulase. *Biotechnol. Biofuels* **2016**, *9*, 70. [CrossRef] [PubMed]
3. Singh, N.; Patel, N.; Singh, J.; Raja, P.; Soni, P.; Parihar, J.S. Carbon exchange in some invasive species in the Himalayan foothills. *Trop. Ecol.* **2016**, *57*, 263–270.
4. Vardien, W.; Richardson, D.M.; Foxcroft, L.C.; Thompson, G.D.; Wilson, J.R.U.; Le Roux, J.J. Invasion dynamics of *Lantana camara* L. (sensulato) in South Africa. *S. Afr. J. Bot.* **2012**, *81*, 81–94. [CrossRef]
5. Ruwanza, S. Soil seed bank depletion as a mechanism of *Lantana camara* L. invasion. *S. Afr. J. Plant Soil* **2016**, *33*, 303–308. [CrossRef]
6. Day, M.D.; Wiley, C.J.; Playford, J.; Zalucki, M.P. *Lantana: Current Management Status and Future Prospects*; (No. 435-2016-33733); Australian Centre for International Agricultural Research: Canberra, Australia, 2003.
7. Sharma, O.P.; Sharma, S.; Pattabhi, V.; Mahato, S.B.; Sharma, P.D. A review of the hepatotoxic plant *Lantana camara*. *Crit. Rev. Toxicol.* **2007**, *37*, 313–352. [CrossRef] [PubMed]
8. Bevilacqua, A.H.V.; Suffredini, I.B.; Romoff, P.; Lago, J.H.G.; Bernardi, M.M. Toxicity of apolar and polar *Lantana camara* L. crude extracts in mice. *Res. Vet. Sci.* **2011**, *90*, 106–115. [CrossRef] [PubMed]
9. Pereira, K.L.G.; Nogueira, P.C.D.L.; Arrigoni-Blank, M.D.F.; Nizio, D.A.D.C.; Silva, D.C.; Pinto, J.A.O.; Blank, A.F. Chemical diversity of essential oils of *Lantana camara* L. native populations. *J. Essent. Oil Res.* **2019**, *32*, 32–47. [CrossRef]
10. Qureshi, H.; Anwar, T.; Ali, Q.; Haider, Z.; Habib, N.; Fatima, S.; Adkins, S.W. Isolation of natural herbicidal compound from *Lantana camara*. *Int. J. Environ. Anal. Chem.* **2019**, *101*, 631–638. [CrossRef]
11. Bhagwat, S.A.; Breman, E.; Thekaekara, T.; Thornton, T.F.; Willis, K.J. A battle lost? Report on two centuries of invasion and management of *Lantana camara* L. in Australia, India and South Africa. *PLoS ONE* **2012**, *7*, e32407. [CrossRef]
12. Sharma, P. An overview of the field of family business studies: Current status and directions for the future. *Fam. Bus. Rev.* **2004**, *17*, 1–36. [CrossRef]
13. Pasha, M.A.; Jayashankara, V.P. p- Toluenesulfonic acid (p- TSA)- catalyzed efficient synthesis of bis (indolyl) methanes under grindstone method. *J. Pharmacol. Toxicol.* **2006**, *1*, 585–590. [CrossRef]
14. Varshney, V.K.; Gupta, P.K.; Naithani, S.; Khullar, R.; Bhatt, A.; Soni, P.L. Carboxymethylation of α-cellulose isolated from *Lantana camara* with respect to degree of substitution and rheological behavior. *Carbohydr. Polym.* **2006**, *63*, 40–45. [CrossRef]
15. Liebeke, M.; Strittmatter, N.; Fearn, S.; Morgan, A.J.; Kille, P.; Fuchser, J.; Wallis, D.; Palchykov, V.; Robertson, J.; Lahive, E.; et al. Unique metabolites protect earthworms against plant polyphenols. *Nat. Commun.* **2015**, *6*, 7869. [CrossRef]
16. Abbasi, S.A.; Nayeem-Shah, M.; Abbasi, T. Vermicomposting of phytomass: Limitations of the past approaches and the emerging directions. *J. Clean. Prod.* **2015**, *93*, 103–114. [CrossRef]
17. Hussain, N. Gainful Utilization of Toxic and Allelopathic Weeds for the Generation of Biofertilizer through Vermicomposting: An Assessment of the Fertilizer Value of the Resultant Vermicompost. Ph.D. Thesis, Pondicherry University, Puducherry, India, 2016; 175p.
18. Hussain, N.; Abbasi, S.A. Efficacy of the vermicomposts of different organic wastes as "clean" fertilizers: State-of-the-art. *Sustainability* **2018**, *10*, 1205. [CrossRef]
19. Hussain, N.; Abbasi, T.; Abbasi, S.A. Vermicomposting eliminates the toxicity of Lantana (*Lantana camara*) and turns it into a plant friendly organic fertilizer. *J. Hazard. Mater.* **2015**, *298*, 46–57. [CrossRef] [PubMed]

20. Abbasi, S.A.; Hussain, N.; Tauseef, S.M.; Abbasi, T. A novel FLippable Units Vermireactor Train System—FLUVTS–For rapidly vermicomposting paper waste to an organic fertilizer. *J. Clean. Prod.* **2018**, *198*, 917–930. [CrossRef]
21. Available online: https://www.krishisewa.com/miscellaneous/soil-fertility/1171-best-use-of-organic-waste-through-vermicomposting.html (accessed on 9 December 2021).
22. Moran, R.; Porath, D. Chlorophyll determination in intact tissues using N, N-dimethylformamide. *Plant Physiol.* **1980**, *65*, 478–479. [CrossRef] [PubMed]
23. Wellburn, A.R. The spectral determination of chlorophylls a and b, as well as total carotenoids, using various solvents with spectrophotometers of different resolution. *J. Plant Physiol.* **1994**, *144*, 307–313. [CrossRef]
24. Nielsen, S.S. (Ed.) *Food Analysis*; Springer: New York, NY, USA, 2010; 550p.
25. Carter, M.R.; Gregorich, E.G. *Soil Sampling and Methods of Analysis*; Canadian Society of Soil Science; Taylor and Francis Group, LLC: Boca Raton, FL, USA, 2007; Volume 2, pp. 743–760.
26. Bashour, I.I.; Sayegh, A.H. *Methods of Analysis for Soils of Arid and Semi-Arid Regions*; Food and Agriculture Organization of the United Nations: Rome, Italy, 2007.
27. Heanes, D.L. Determination of total organic-C in soils by an improved chromic acid digestion and spectrophotometric procedure. *Commun. Soil Sci. Plant Anal.* **1984**, *15*, 1191–1213. [CrossRef]
28. Kandeler, E. Bestimung von Gesamtstickstoffnachkjeldahl. In *Bodenbiologische Arbeitsmethoden*; Schinner, F., Kandeler, E., Ohlinger, R., Margesin, R., Eds.; Springer: Berlin, Germany, 1993; pp. 346–366.
29. Cho, J.T.; Moon, B.J. The occurrence of strawberry black leaf spot caused by *Alternaria alternata* (Fr.) Keissler in Korea. *Korean J. Plant Prot.* **1980**, *19*, 221–227.
30. Ahmed, M.M.M. Studies on the Control of Insect Pests in Vegetables (Okra, Tomato, and Onion) in Sudan with Special Reference to Neem-Preparations. Ph.D. Thesis, University of Giessen, Giessen, Germany, 2000; p. 112.
31. Hussain, N.; Abbasi, T.; Abbasi, S.A. Generation of highly potent organic fertilizer from pernicious aquatic weed *Salvinia molesta*. *Environ. Sci. Pollut. Res.* **2018**, *25*, 4989–5002. [CrossRef] [PubMed]
32. Hilhorst, H.W.; Karssen, C.M. Effect of chemical environment on seed germination. In *Seeds. The Ecology of Regeneration in Plants Communities*; Cab International: Oxon, UK, 2000; pp. 293–309.
33. Ievinsh, G. Vermicompost treatment differentially affects seed germination, seedling growth and physiological status of vegetable crop species. *Plant Growth Regul.* **2011**, *65*, 169–181. [CrossRef]
34. Edwards, C.A.; Norman, Q.A.; Sherman, R. *Vermiculture Technology, Earthworms Organic Waste and Environmental Management*; CRC Press: Boca Raton, FL, USA, 2011; pp. 220–231.
35. Roberts, P.; Edwards, C.A.; Edwards-Jones, G.; Jones, D.L. Responses of common pot grown flower species to commercial plant growth media substituted with vermicomposts. *Compost Sci. Util.* **2007**, *15*, 159–166. [CrossRef]
36. Doan, T.T.; Henry-des-Tureaux, T.; Rumpel, C.; Janeau, J.L.; Jouquet, P. Impact of compost, vermicompost and biochar on soil fertility, maize yield and soil erosion in Northern Vietnam: A three-year mesocosm experiment. *Sci. Total Environ.* **2015**, *514*, 147–154. [CrossRef]
37. Edwards, C.A.; Arancon, N.Q.; Bennett, M.V.; Askar, A.; Keeney, G.; Little, B. Suppression of green peach aphid (*Myzuspersicae*) (Sulz.), citrus mealybug (*Planococcuscitri*), and two spotted spider mite (*Tetranychusurticae*) (Koch.) attacks on tomatoes and cucumbers by aqueous extracts from vermicomposts. *Crop Prot.* **2010**, *29*, 80–93. [CrossRef]

agriculture

MDPI

Article

Using Biochar and Vermiwash to Improve Biological Activities of Soil

Michelangelo Becagli, Iduna Arduini * and Roberto Cardelli

Department of Agriculture, Food and Environment, University of Pisa, 56124 Pisa, Italy; michelangelo.becagli@phd.unipi.it (M.B.); roberto.cardelli@unipi.it (R.C.)
* Correspondence: iduna.arduini@unipi.it; Tel.: +39-050-2218902

Abstract: The recycling of key nutrients and bioenergy from waste materials is a goal of sustainable agriculture. The co-application of biochar and a vermicompost solution (vermiwash) could enhance the positive effects of both materials on soil biomass and biological activities. Tomato plants were grown in soil amended with biochar, mixed at a rate of 2% w/w, and vermiwash, applied through fertigation at a rate of 25 mg per plant, alone (B and V) and in combination (BV). Organic C, dissolved organic C (DOC), soil biomass C, and some enzymatic activities were determined at the start (T0) and the end (T100) of the cultivation period in bulk soil and rhizosphere soil. B and V significantly increased the organic C and soil biomass contents. In addition, B retained the DOC species derived from the soil and, in the BV treatment, also the humic substance of the vermiwash. Generally, all the parameters achieved higher values in the rhizosphere than in the bulk soil. The altered soil index three (AI3) of enzyme activities suggests that applying V and B is helpful for the soil microorganisms. Synergisms between B and V were low in the bulk soil and clearly evident in the rhizosphere.

Keywords: biochar; β-glucosidase; organic fertilisers; phosphatase; soil enzyme activity; soil microbiome; *Solanum lycopersicum*; urease; vermicompost

1. Introduction

Biochar is a pyrolysis product derived from the thermal decomposition in the absence of oxygen of various organic, primarily woody, sources. It has an aromatic structure, is rich in carbon, and has stable physical and chemical properties. Biochar improves the productivity of cultivated plants not only because of its nutrient content but also through better nutrient retention, increase in soil cation exchange capacity, and an improvement in the physical properties of the soil, including an increase in water retention [1].

Many studies indicate that biochar can increase the biomass of soil microorganisms and stimulate their enzymatic activity [2]. However, the addition of biochar to soil has also led to conflicting data on the activity of given enzymes [3]. The biochar introduced into the soil creates a favourable habitat for microorganisms, especially in terms of higher soil porosity. Soil enzymes catalyse the biodegradation phases of different substrates, thus favouring their decomposition. The complex factors influencing soil biology are very wide, and therefore evaluating the enzymatic activities is a necessary step towards a complete understanding of the key processes that connect populations of microorganisms and trace element dynamics, following the application of biochar to soils [4].

Biochar can be enriched with organic and/or mineral nutrients to mitigate its possible negative effects and the addition of compost was found to increase the organic C of the soil and influence the enzyme activity [5,6]. Vermicomposts are generally more stable than composts, have greater availability of mineral nutrients, and better biological properties [7]. Tejada and González [8] demonstrated that vermicompost improved both soil biomass and the activity of selected soil enzymes, which increased the yield and quality of rice crops. Earthworms play a fundamental role in vermicomposting together with microbes in the conversion of solid organic waste into a soil conditioner that is rich in nutrients. During the production of vermicompost, various products are created such as vermicompost tea, vermi-liquid (liquid collected during the worming process), and earthworm biomass. The leachate (vermiwash) is generated together with the vermicomposting process, commonly referred to as vermicomposting leachate or worm-bed leachate [9]. This product contains soluble nutrients in addition to various organic acids and mucus from earthworms and microbes [10]. The novelty of this work is the co-application of vermiwash and biochar to soil, hypothesising that activating biochar with vermiwash will improve the biological activity in the soil. By recovering and recycling key nutrients and bioenergy from waste materials, this approach reduces the use of inorganic fertilisers and, thus, meets the objectives of sustainable agriculture which are the subject of the Farm to Fork strategy of the European Union. Moreover, it applies agroecological farming methods that are growing in global interest as attractive and safer alternatives to the use of imported fertilisers and chemicals for food production [11].

This study aimed to evaluate the impact of vermiwash, biochar, and their combination on the organic substance and biological activities of a soil used for tomato cultivation in greenhouse conditions. The study proved that both materials increased soil fertility and stimulated rhizosphere activity.

2. Materials and Methods

2.1. Experiment Setup

A randomised experiment consisting of four treatments with five replicates was conducted in a greenhouse at the Department of Agriculture, Food and Environment of the University of Pisa. Treatments consisted of a control soil without amendments (CTR), soil amended once a week with vermiwash (V), soil amended with biochar obtained by pyrogasification (B), and soil amended with the same biochar treatment and periodically treated as for V (BV).

2.2. Materials

Agricultural soil was collected at a depth between 0 and 15 cm from a field at the agricultural research centre "Enrico Avanzi" of the University of Pisa (Lat. 43°39′38.96″ N; Long. 10°18′22.17″ E; 1 m above sea level) and, after being air dried, it was sieved through a 2 mm mesh to remove large fragments. The soil was classified as sandy, Typic Xerorthent. The main soil properties were determined following standard methods [12]: 86.3% sand, 7.9% silt, 5.8% clay, 48% water holding capacity (WHC), 8.2 pH, 5.8% $CaCO_3$, 0.91% organic C, 0.15 g kg^{-1} DOC, 1.01 g kg^{-1} total N, 10.4 mg/kg available p, 78.6 mg/kg exchangeable K, and 4.93 $cmol^{(+)}$/kg cation exchange capacity.

Biochar was produced from woodchips (30–50 g) of pristine forests (*Abies* sp., *Fagus* sp., *Robinia pseudoacacia*) by pyrogasification with a co-current fixed bed ("downdraft") "syngaSmart®" gasifier (RESET s.r.l. https://www.reset-energy.com/; accessed on 24 January 2022). The average heating rate before reaching a peak of 800 °C was 15 °C to 18 °C/min. The characterisation of biochar followed the official methods approved by Italian regulations (D.lgs. 75/2015), as reported in Table 1. Following the Guidelines for Certification of the International Biochar Initiative (IBI, http://www.european-biochar.org/en/ebc-ibi; accessed on 24 January 2022), the total organic carbon content was classified as Class 1. The biochar was applied to soil at a rate of 2% *w*/*w*, which corresponded to 34 t/ha in dry weight, and the physical and chemical parameters of the mixture were analysed (Table 2).

Table 1. Selected characteristics of the biochar used.

Parameter	U.m.	Biochar
Water Holding Capacity	%	400
pH		11.3
Total C	%	68.5
Organic C	%	68.4
Total N	g/kg	0.516
C:N ratio		132.7
Total P	mg/kg	340
Total K	mg/kg	4.3

Table 2. Soil characteristics after amendment with biochar.

Parameter	U.m.	Soil and Biochar (2% *w/w*)
Water Holding Capacity	%	52
pH		8.4
Organic C	%	3.4
Cation Exchange Capacity	$cmol^{(+)}$/kg	9.54
Total N	g/kg	1.02
Available *p*	mg/kg	17.2
Exchangeable K	mg/kg	164.6
Water Holding Capacity	%	52

The vermicompost used in this experiment was obtained from mature pig manure derived from organic livestock. The initial biomass was converted into vermicompost in an open-air litter with the earthworm species *Eisenia fetida* and *Eisenia andrei*. The C/N ratio of the mature vermicompost was 14, and the concentration in heavy metals was undetectable or below the limits set by the Italian regulation. After four months, the vermiwash was produced in a laboratory-scale plant through cold extraction, which means that a surplus of cold water was applied to vermicompost, and the leachates were collected. The chemical properties of vermiwash were: 60% organic matter, 80% total humic substances (in organic matter content), pH 8.00, 2% total nitrogen content, 1.5% total organic nitrogen, and 20 C/N ratio. The microbial characterisation is reported in Table 3. Vermiwash was applied weekly through fertigation at a rate of 25 mg per pot [13], corresponding to 300 g on 1000 m^2.

Table 3. Microbiological composition of vermiwash. CFU, colony forming unit; MPN, most probable number.

Functional Diversity Group	U.m.	Composition
Amylolytic	CFU/mL	2.2×10^3
Cellulolytic	CFU/mL	1.4×10^4
Nitrosant	MPN/mL	2×10^2
Nitricant	MPN/mL	4.5×10^3
Sulphate reducers	MPN/mL	4.5×10^2
Sulphur oxidants	MPN/mL	2.5×10^1
Aerobic nitrogen-fixing	MPN/mL	2.5×10^3
Anaerobic nitrogen-fixing	MPN/mL	2×10^3

2.3. Experiment Management

The mixture of soil and biochar for B and BV treatments was prepared one week before filling the pots, and vermiwash solution was added in V and BV and all pots were wetted at 60% WHC to enable the biochar to be colonised by the microorganisms of soil and V. Tomato plants (*Solanum lycopersicum* L.) obtained from a commercial nursery were individually transplanted at the 3-leaf stage (30 DAS) into 18 L plastic pots containing 16 kg of soil.

To ensure an adequate nitrogen, phosphorus, and potassium nutrition for the plants, a basal dressing was prepared for each pot with potassium nitrate and calcium nitrate (60 N kg/ha), monopotassium phosphate (30 P_2O_5 kg/ha) and potassium sulphate (120 K_2O kg/ha). The remaining nutritional requirements of the recommended doses for tomato (180 N kg/ha; 30 P_2O_5 kg/ha; 120 K_2O kg/ha) were provided through fertigation once a week.

2.4. Chemical Analyses

Soil samples were collected at seedling planting (T0) and after 100 days, corresponding to the stage of fruit ripening. At T100, rhizosphere (Rz) and bulk (Bk) soil fractions were obtained following the method described in Barillot et al. [14].

Total organic carbon (TOC) was determined by dry combustion (induction furnace 900 CS, Eltra), after the removal of carbonate C. Dissolved organic carbon (DOC) was determined with an organic C analyser for liquid samples (Hach QbD1200) after stirring soil samples with distilled water (soil/H_2O 1:20) at room temperature for 24 h, centrifuging the suspension at 10,000 rpm for 10 min and, then, filtering it through a 0.45 mm glass fibre. The soil microbial biomass C (MBC) was determined following the method of Vance et al. [15], which consisted of the extraction of organic C from fumigated and unfumigated soils with a 1 N K_2SO_4 solution. The obtained organic C was measured with a QBD1200 Laboratory TOC Analyser (Hach Company, Loveland, CO, USA). To convert into microbial biomass C the difference in soluble C between the fumigated and unfumigated soils, an extraction efficiency coefficient (Kc) equal to 0.45 was used.

The enzymatic analyses were performed as described by Cardelli et al. [16]. Dehydrogenase activity (DH) was estimated with the colorimetric assay of the 2,3,5-triphenylformazan (TPF) obtained by the microorganism from the reduction of 2,3,5-triphenyltetrazolium chloride (TTC). To estimate the β-glucosidase activity (GL), soil samples were incubated at 37 °C for 60 min and the reaction product p-nitrophenol obtained from the substrate 4-nitrophenyl-ß-D-glucopiranosyde was determined colorimetrically, at 410 nm. To determine the alkaline phosphatase activity (AP), p-nitrophenyl phosphate was added to the soil samples and the p-nitrophenol released by hydrolysis was measured colorimetrically. To estimate the arylsulphatase activity (AS), soil samples were incubated at 37 °C for 1 h with the substrate p-nitrophenyl sulphate. The p-nitrophenol produced by the microorganisms was extracted by dilute alkali ($CaCl_2$ 0.5M and NaOH 0.5M) and determined at 400 nm with a colorimetric method. Urease activity (UR) was assessed by a spectrophotometric method measuring the ammonia released from urea after a 2 h incubation of soil samples with urea at 37 °C.

The alteration index (AI3) combines the activity of three enzymes to estimate the degree of soil quality alteration in response to treatments. Following Puglisi et al. [17], AI3 was calculated as:

$$AI3 = (7.87 \times \beta\text{-glucosidase}) - (8.22 \times \text{phosphatase}) - (0.49 \times \text{urease})$$

where the activities of enzymes were expressed in micromoles of p-nitrophenol per gram of soil per hour (for β-glucosidase and phosphatase), and in micrograms of urea per gram of soil per hour (urease).

2.5. Statistical Analysis

ANOVA was used to analyse the effects of the treatment sample time (T0 and T100), fertiliser (CTR, V, B, BV), soil region (Bk, Rz), and their interactions, with data arranged in a split-split plot design with five replicates, each consisting of one pot for one tomato plant. The JMP software (SAS Institute, Inc., Cary, NC, USA) was used, and the Tukey–Kramer post-hoc test was used to separate significantly different means using $p < 0.05$.

3. Results and Discussion

3.1. Total and Dissolved Organic C

As expected, the addition of vermiwash and especially biochar to the soil increased the total organic carbon (TOC) content at T0 (Figure 1). In all treatments, tomato cultivation did not significantly affect TOC in the bulk soil, while rhizosphere soil values were approximately 40% higher in the presence of B. However, in the Bk soil at T100, 88% of the initial TOC was found in the control, whereas 92% and 94% were found in V and BV, respectively, and only 80% in B, thus indicating a faster mineralisation of the soil organic matter. These findings agree with Awad et al. [18], who found that the decomposition of plant residues in soil was enhanced by biochar, probably because the higher soil aeration and porosity induced by biochar favoured microbial growth and respiration [19]. A further explanation could be a positive priming effect [20]. The % of organic C was significantly higher in the rhizosphere soil than in bulk soil in all the treatments, to which probably contributed root exudates in the form of easily decomposable polysaccharides (O/N-alkyl C) [21].

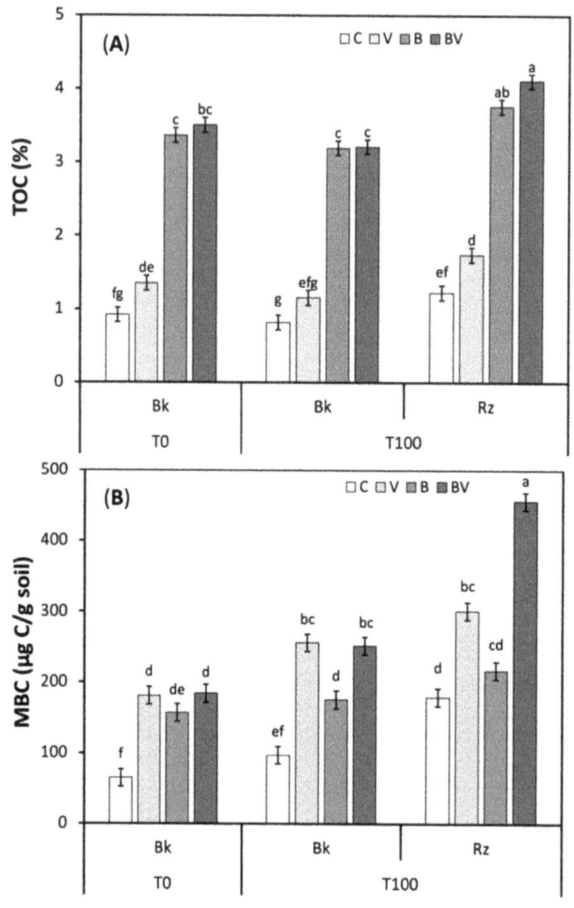

Figure 1. Total organic C (**A**) and dissolved organic C (**B**) in the bulk (Bk) and rhizosphere (Rz) soils amended with different fertilisers at the start (T0) and end (T100) of tomato cultivation. CTR—control; V—vermiwash; B—biochar; BV—biochar together with vermiwash. Bars represent SE of the interaction sample time × fertiliser × soil region. Different letters indicate significant differences among treatments ($p < 0.05$; Tukey's test).

At T0, the dissolved organic carbon (DOC) was highest in the soil treated with vermiwash and lowest when containing biochar, demonstrating that the labile organic matter of V was promptly stabilised by biochar, as reported by Schulz and Glaser [22] for compost (Figure 1). After 100 days in the Bk soil, the DOC decreased in CTR and V, and increased in B and especially in BV. The decrease with time in CTR and V could be due in part to the leaching caused by irrigation, and in part to the faster degradation of water-soluble C that is in the first stage of mineralisation [23]. Biochar characteristics, such as high or low pyrolysis temperature, typologies (herbaceous, woody, or animal-derived by-products), biochar/soil ratio, and soil characteristics (texture, pH, CEC, nutrient content) may affect the impact of biochar on DOC [24]. In the Bk soil, biochar retained the humified DOC species derived from the soil, but also the humic substance present in the vermiwash of the BV treatment, thus proving to be an essential instrument for preserving the soil organic matter (Figure 1). Our results could be explained by the selective behaviour of the pyroligneous carbon of biochar in retaining humic substances compared to labile organic compounds [25,26]. In the Rz soil, trends of DOC were similar as in the bulk soil in response to fertiliser, but in CTR and V, the decrease over time was lower than in the Bk soil, whereas in B and especially BV, the increase was more pronounced, so that DOC values were significantly higher for all treatments (Figure 1). These patterns suggest that root exudates contributed to increase DOC in the Rz soil and that the release was stimulated by vermiwash.

3.2. Soil Biomass

The incorporation of fertilisers in the soil increased the microbial biomass C (MBC) from 65 μg/g in CTR, to 157 μg/g in B, and to approximately 183 μg/g in V and BV (Figure 2), which was due to the stimulation of soil microbiota in response to the easily available C and/or to the addition of foreign microorganisms with the materials [26,27].

Higher MBC was found in vermiwash compared to compost [28], and Uz et al. [29] also reported a strong increase in bacterial number when V was added to an alkaline soil. The MBC tended to increase over time in all treatments, but the increments were significant for all soil regions in V and BV, only for the Rz soil in the control, and never for B, probably because the addition of biochar only moderately increased soil DOC, which is the most available substrate for microbial growth [30]. In the Bk soil, the high MBC recorded at T100 in the V and BV treatments was largely due to the weekly soil fertilisation, which undoubtedly stimulated microbial proliferation. At T100, the MBC of the BV treatments were approximately 80% higher in the Rz than in the Bk soil, while the increments were by only 20% in V and B, thus demonstrating a strong interaction of the two fertilisers in the rhizosphere, which affected positively tomato growth [31]. Gopal et al. [32] hypothesised that coconut leaf vermiwash, a wash of composting substrates and earthworm bodies, led to increased microbial populations by promoting soil nutrient content, and acted as a liquid fertiliser immediately and quickly absorbed by the roots of plants; in addition, the vermiwash by itself showed a very low microbial population, and on application to soil, the soil microbial population increased. In the Rz soil, the MBC of CTR reached similar values to B, revealing a key role of root exudates in stimulating microbial growth in unfertilised soils. The expression of microbial C as a percentage of organic C highlighted that vermiwash alone greatly stimulated the microbial biomass, with the highest values recorded in the Bk soil at T100 (Figure 2). The patterns of MBC% observed in the Rz suggest a higher rhizodeposition of C compounds in both CTR- and vermiwash-amended soils.

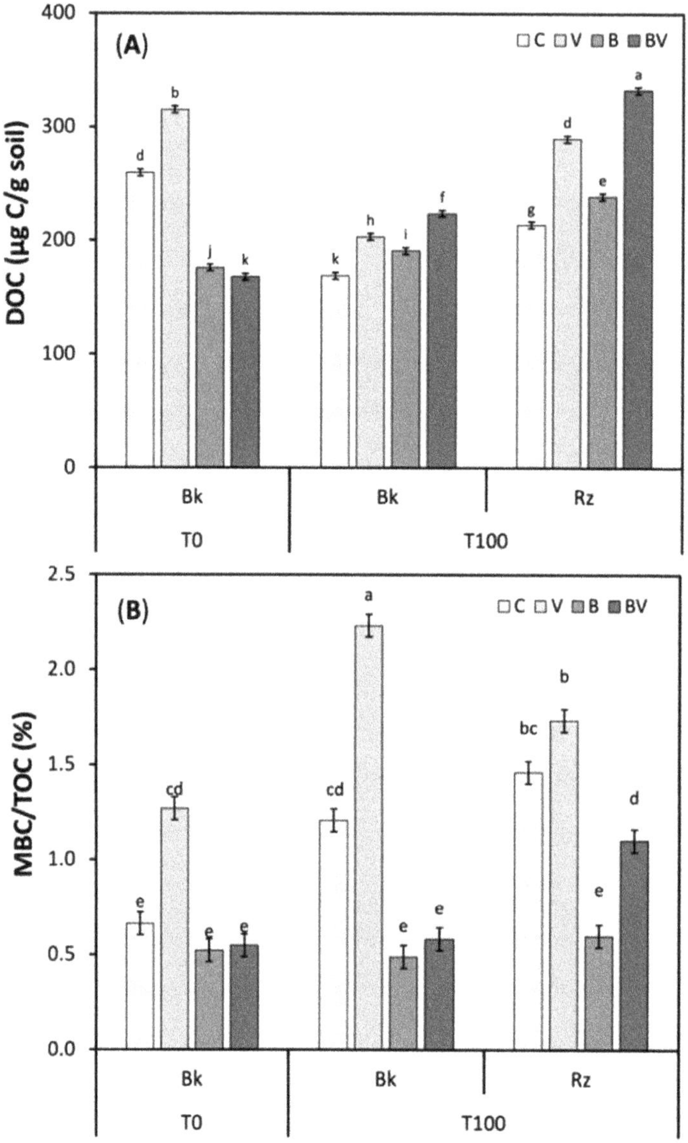

Figure 2. Microbial biomass C (**A**) and MBC percentage of total organic C (**B**) in the bulk (Bk) and rhizosphere (Rz) soils amended with different fertilisers at the start (T0) and end (T100) of tomato cultivation. CTR—control; V—vermiwash; B—biochar; BV—biochar together with vermiwash. Bars represent SE of the interaction sample time × fertiliser × soil region. Different letters indicate significant differences among treatments ($p < 0.05$; Tukey's test).

3.3. Soil Enzymatic Activities

The addition of biochar and vermiwash to soil stimulated all the enzymatic activities tested in this study and, in general, activities increased after tomato cultivation and tended to be higher in the rhizosphere than the bulk soil (Table 4).

Table 4. Enzyme activities of bulk and rhizosphere soils amended with different fertilisers at the start (T0) and end (T100) of tomato cultivation. V—vermiwash; B—biochar; BV—biochar together with vermiwash.

	T0	T100	
Fertiliser	**Bulk**	**Bulk**	**Rhizosphere**
		Dehydrogenase (μg TTF/g·h)	
CTR	0.22 ± 0.007 c,d	0.38 ± 0.012 c,d	1.10 ± 0.040 a,b
V	0.19 ± 0.005 c,d	0.75 ± 0.034 b,c	1.63 ± 0.060 a
B	0.20 ± 0.004 c,d	0.52 ± 0.019 c,d	1.62 ± 0.050 a
BV	0.12 ± 0.005 d	0.46 ± 0.021 c,d	1.61 ± 0.050 a
		Phosphatase (μg p-nitrophenol/g·h)	
CTR	76.6 ± 2.80 h	118.9 ± 2.50 f,g	172.6 ± 3.44 c
V	144.2 ± 3.60 d,e	141.0 ± 4.03 d,e,f	202.0 ± 5.66 b
B	123.0 ± 3.10 e,f	143.0 ± 4.30 d,e,f	250.5 ± 7.80 a
BV	97.7 ± 2.90 g,h	150.3 ± 4.75 c,d	236.2 ± 8.20 a
		B-glucosidase (μg p-nitrophenol/g·h)	
CTR	5.0 ± 0.19 g	32.5 ± 0.67 e,f	78.4 ± 1.81 c
V	15.7 ± 0.51 f,g	68.2 ± 1.69 c,d	108.3 ± 2.94 a,b
B	24.9 ± 0.46 f,g	51.4 ± 1.29 d,e	86.3 ± 1.89 b,c
BV	22.5 ± 0.47 f,g	51.3 ± 0.95 d,e	124.1 ± 3.01 a
		Arylsulphatase (μg p-nitrophenol/g·h)	
CTR	0.5 ± 0.02 f	4.9 ± 0.12 d,e	9.2 ± 0.45 b,c
V	1.3 ± 0.06 f	10.0 ± 0.39 a,b,c	12.3 ± 0.57 a
B	1.7 ± 0.10 f	7.3 ± 0.34 c,d	10.5 ± 0.21 a,b
BV	2.4 ± 0.09 e,f	7.4 ± 0.31 c,d	11.0 ± 0.27 a,b
		Urease (mg NH_4^+-N/g·2h)	
CTR	37.3 ± 1.40 e	112.7 ± 4.34 d,e	356.0 ± 15.4 c
V	30.3 ± 1.30 e	158.4 ± 6.68 d	630.6 ± 28.1 a
B	79.3 ± 2.20 e,d	144.4 ± 5.89 d	467.7 ± 22.2 b
BV	55.3 ± 1.60 e	153.0 ± 5.57 d	652.2 ± 24.8 a

Values are means ± SE; $n = 5$. Different letters indicate significant differences among means ($p < 0.05$; Tukey's test).

At T100, DH-ase activity did not differ significantly among fertiliser treatments within soil regions, but it was markedly higher in the Rz than the Bk soil: 2.2 times higher in V, approximately 3-fold in CTR and B, and 3.5 times higher in BV. In the bulk soil, however, DH-ase activity was slightly higher in V (0.75 μg/g/h TTF) than in B and BV (0.52 and 0.46 μg/g/h TTF, respectively), while in the Rz soil, it was quite similar in all treatments (1.62 μg/g/h TTF on average), suggesting that root exudates counterbalanced the negative effects of biochar. High soil DH-ase activity following V applications was reported by Arancon et al. [33], whereas the lower DH-ase found in BV compared with V can be explained by the presence of various toxic compounds in the biochar such as polycyclic aromatic hydrocarbons and volatile organic compounds [34]. Although losses of DH-ase in mixtures could be attributed to the decreasing effects of B on the enzyme activity, the values may also be underestimated because of the impact of biochar on assay constituents [16]. The lower level of DH-ase activity in the B treatments could also be explained by the findings of Swaine et al. [35], who reported that biochar amendments led to a significant reduction in concentrations of substrate and extractable product in soil DH-ase assays, thus limiting the identification of biochar effects on soil enzyme activity.

At T0, the alkaline phosphatase activity was significantly higher than in CTR in V and B, but not in the combined application (BV), in which it even seemed to be depressed (Table 4). After tomato cultivation, AP-ase activity increased in both soil regions in the CTR and BV treatments, only in the Rz in V and B, and, for all treatments, values were significantly higher in Rz than in Bk at T100. Positive effects of biochar and vermiwash on AP-ase activity were reported by Lehmann et al. [26] and Uz et al. [29], but our results showed that the combined addition of the two materials did not enhance AP-ase activity in soil, which tended to be higher in the presence of biochar, especially in the rhizosphere (Table 4).

The activities of β-glucosidase and arylsulphatase were undoubtedly stimulated by the presence of V and B, and for both, the differences to CTR became significant only after plant cultivation and were greatest with V in the Bk soil and with both V and BV in the Rz soil (Table 4). An increase in both GL and AS activities with the application of biochar to the soil were reported by Luo and Gu [36] and Lu et al. [37], whereas Günal et al. [38] found that β-glucosidase enzyme activity was found to vary greatly in response to the type of soil and biochar, and to the rates of application of biochar and fertilisers. In addition, Lim et al. [39] found that vermicompost increased β-glucosidase and other hydrolytic enzymes that play key roles in C, P, and S cycling.

The effect of V and B additions on urease activity increased considerably during tomato cultivation, but the differences to CTR became significant only in the rhizosphere soil, with approximately 82% higher UR activity in V and BV and 31% higher in B (Table 4). Among all tested enzyme activities, UR showed the greatest differences between soil regions, being approximately four times higher in the rhizosphere of tomato compared to the bulk soil, with a BV > V > B = C trend. A surprisingly higher urease activity in the Rz compared to the Bk soil, associated with increasing variations in the fungal community during plant growth, was also observed after the shift to conservation tillage, and was imputed to changes in the soil nutrient status favoured by an increase in root exudates [40].

The alteration index three (AI3) is a data reduction process that involves the activities of three key enzymes: β-glucosidase, phosphatase, and urease, that are converted into scores reflecting the degree of positive or negative changes in the soil (alteration). The AI3 may be negative and positive and does not have target values [17]. Meyer et al. [41] reported that AI3 was correlated with soil organic matter content and yield performance. Analysing a set of amended and unamended soils, Puglisi et al. [17] observed that soils with more negative AI3 values had higher total organic carbon (TOC) content. Several studies confirmed the tendency of AI3 scores to become increasingly negative with increasing soil organic carbon content and soil quality [42,43].

In this study, we used AI3 to compare altered (V, B, and BV) and unaltered (CTR) soils. In all treatments, the AI3 values became more negative from T0 to T100, thus highlighting an amelioration of soil quality during tomato cultivation (Table 5). In the bulk soil, AI3 did not differ significantly among fertiliser treatments at both T0 and T100, whereas the Rz values were more negative in the fertilised soils and especially in those amended with vermiwash.

Table 5. Alteration index three (AI3) of bulk and rhizosphere soils amended with different fertilisers at the start (T0) and end (T100) of tomato cultivation. CTR—control; V—vermiwash; B—biochar; BV—biochar together with vermiwash.

	T0	**T100**	
Fertiliser	**Bulk**	**Bulk**	**Rhizosphere**
CTR	-7.4 ± 0.23 [a]	-12.9 ± 0.45 [bc]	-21.6 ± 0.95 [d]
V	-12.2 ± 0.41 [a,b,c]	-13.0 ± 0.54 [b,c]	-32.8 ± 1.11 [e]
B	-10.9 ± 0.28 [a,b,c]	-14.5 ± 0.63 [c]	-29.4 ± 0.98 [d]
BV	-8.2 ± 0.24 [a,b]	-15.6 ± 0.65 [c]	-34.0 ± 1.14 [f]

Values are means ± SE; $n = 5$. Different letters indicate significant differences among means ($p < 0.05$; Tukey's test).

These patterns demonstrate: (i) that root exudates play a key role in supporting soil biota, and (ii) that the application of appropriate amounts of V and B is helpful for the microorganisms of soil, leading to a higher biological quality of soil in the root environment. Moreover, the more negative AI3 recorded in BV at T100, both in the Bk and Rz soil, highlights that biochar is able to absorb and retain the organic molecules and the humic acids present in vermiwash, thus reducing both carbon and nitrate losses [31].

4. Conclusions

Both biochar and vermiwash provide the soil with organic matter and increase the microbial biomass, thus demonstrating to be useful to increase the native soil fertility. Soil enzymatic activities were stimulated by the presence of the two materials and increased after tomato cultivation, with slight synergies between biochar and vermiwash. The ranking among fertiliser treatments was similar in the two soil regions, but all enzyme activities turned out to be higher in the rhizosphere, revealing the contribution of root exudates to soil C metabolism.

The combined application of biochar and vermiwash could be a strategic and environmentally friendly instrument for preserving soil quality and reducing C losses in sustainable agriculture.

Further investigations should address the behaviour of biochar and vermiwash in the medium–long term, and in nutrient-poor soils.

Author Contributions: All authors contributed to planning the experiment, and to the elaboration of the data and the preparation of the manuscript. Specific focus: conceptualisation, R.C. and M.B.; methodology, R.C., M.B., and I.A.; investigation, M.B.; data curation, R.C. and M.B.; writing—original draft preparation, M.B.; writing—review and editing, I.A.; visualisation and supervision, R.C. All authors have read and agreed to the published version of the manuscript.

Funding: This research received no external funding.

Institutional Review Board Statement: Not applicable.

Informed Consent Statement: Not applicable.

Data Availability Statement: The raw data supporting the conclusions of this manuscript are made available by the authors to any qualified researcher who makes a request.

Acknowledgments: The authors acknowledge RESET s.r.l. (Renewable Energy Solutions Environmental Technology, Rieti, Italy) for having produced and provided the biochar.

Conflicts of Interest: The authors declare no conflict of interest.

References

1. Laird, D.A.; Fleming, P.; Davis, D.D.; Horton, R.; Wang, B.; Karlen, D.L. Impact of biochar amendments on the quality of a typical Midwestern agricultural soil. *Geoderma* **2010**, *158*, 443–449. [CrossRef]
2. Lehmann, J.; Joseph, S. *Biochar for Environmental Management. Science, Technology and Implementation*, 2nd ed.; Routledge: London, UK, 2015; p. 944.
3. Mierzwa-Hersztek, M.; Gondek, K.; Baran, A. Effect of poultry litter biochar on soil enzymatic activity, ecotoxicity and plant growth. *Appl. Soil Ecol.* **2016**, *105*, 144–150. [CrossRef]
4. Ameloot, N.; De Neve, S.; Jegajeevagan, K.; Yildiz, G.; Buchan, D.; Funkuin, Y.N.; Prins, W.; Bouckaert, L.; Sleutel, S. Short-term CO_2 and N_2O emissions and microbial properties of biochar amended sandy loam soils. *Soil Biol. Biochem.* **2013**, *57*, 401–410. [CrossRef]
5. Joseph, S.; van Zwieten, L.; Chia, C.; Kimber, S.; Munroe, P.; Lin, Y.; Marjo, C.; Hook, J.; Thomas, T.; Nielsen, S.; et al. Designing specific biochars to address soil constraints: A developing industry. In *Biochar and Soil Biota*; Ladygina, N., Rineau, F., Eds.; CRC Press: Boca Raton, FL, USA, 2013; pp. 165–201.
6. Cardelli, R.; Becagli, M.; Marchini, F.; Saviozzi, A. Effect of biochar, green compost, and vermicompost on the quality of a calcareous soil: A 1-year laboratory experiment. *Soil Sci.* **2017**, *182*, 248–255. [CrossRef]
7. Pramanik, P.; Ghosh, G.K.; Ghosal, P.K.; Banik, P. Changes in organic—C, N, P and K and enzyme activities in vermicompost of biodegradable organic wastes under liming and microbial inoculants. *Bioresour. Technol.* **2007**, *98*, 2485–2494. [CrossRef]
8. Tejada, M.; González, J.L. Application of two vermicomposts on a rice crop: Effects on soil biological properties and rice quality and yield. *Agron. J.* **2009**, *101*, 336–344. [CrossRef]
9. Gutiérrez-Miceli, F.A.; Llaven, M.A.O.; Nazar, P.M.; Sesma, B.R.; Álvarez-Solís, J.D.; Dendooven, L. Optimization of vermicompost and worm-bed leachate for the organic cultivation of radish. *J. Plant Nutr.* **2011**, *34*, 1642–1653. [CrossRef]
10. Shivsubramanian, K.; Ganeshkumar, M. Influence of vermiwash on biological productivity of Marigold. *Madras Agric. J.* **2004**, *91*, 221–225.
11. Cissé, D.; Cornelis, J.T.; Traoré, M.; Saba, F.; Coulibaly, K.; Lefebvre, D.; Colinet, G.; Nacro, H.B. Co-composted biochar to decrease fertilization rates in cotton-maize rotation in Burkina Faso. *Agron. J.* **2021**, *113*, 5516–5526. [CrossRef]
12. Colombo, C.; Miano, T. *Metodi di Analisi Chimica del Suolo*; Società Italiana della Scienza del Suolo (SISS): Bari, Italy, 2015.

13. Olivares, F.L.; Aguiar, N.O.; Rosa, R.C.C.; Canellas, L.P. Substrate biofortification in combination with foliar sprays of plant growth promoting bacteria and humic substances boosts production of organic tomatoes. *Sci. Hortic.* **2015**, *183*, 100–108. [CrossRef]
14. Barillot, C.D.C.; Sarde, C.O.; Bert, V.; Tarnaud, E.; Cochet, N. A standardized method for the sampling of rhizosphere and rhizoplan soil bacteria associated to a herbaceous root system. *Ann. Microbiol.* **2013**, *63*, 471–476. [CrossRef]
15. Vance, E.D.; Brookes, C.; Jenkinson, D.S. An extraction method for measuring soil microbial biomass carbon. *Soil Biol. Biochem.* **1987**, *19*, 403–707. [CrossRef]
16. Cardelli, R.; Becagli, M.; Marchini, F.; Saviozzi, A. Biochar impact on the estimation of the colorimetric-based enzymatic assays of soil. *Soil Use Manag.* **2019**, *35*, 478–481. [CrossRef]
17. Puglisi, E.; Del Re, A.A.M.; Rao, M.A.; Gianfreda, L. Development and validation of numerical indexes integrating enzyme activities of soils. *Soil Biol. Biochem.* **2006**, *38*, 1673–1681. [CrossRef]
18. Awad, Y.M.; Blagodatskaya, E.; Ok, Y.S.; Kuzyakov, Y. Effects of polyacrylamide, biopolymer, and biochar on decomposition of soil organic matter and plant residues as determined by 14C and enzyme activities. *Eur. J. Soil Biol.* **2012**, *48*, 1–10. [CrossRef]
19. Lei, O.; Zhang, R. Effects of biochars derived from different feedstocks and pyrolysis temperatures on soil physical and hydraulic properties. *J. Soils Sediments* **2013**, *13*, 1561–1572. [CrossRef]
20. Zimmerman, A.R.; Gao, B.; Ahn, M.Y. Positive and negative carbon mineralization priming effects among a variety of biochar-amended soils. *Soil Biol. Biochem.* **2011**, *43*, 1169–1179. [CrossRef]
21. Angst, G.; Messinger, J.; Greiner, M.; Häusler, W.; Hertel, D.; Kirfel, K.; Kögel-Knabner, I.; Leuschner, C.; Rethemeyer, J.; Mueller, C.W. Soil organic carbon stocks in topsoil and subsoil controlled by parent material, carbon input in the rhizosphere, and microbial-derived compounds. *Soil Biol. Biochem.* **2018**, *122*, 19–30. [CrossRef]
22. Schulz, H.; Glaser, B. Effects of biochar compared to organic and inorganic fertilizers on soil quality and plant growth in a greenhouse experiment. *J. Plant Nutr. Soil Sci.* **2012**, *175*, 410–422. [CrossRef]
23. Pascual, J.A.; García, C.; Hernandez, T.; Ayuso, M. Changes in the microbial activity of an arid soil amended with urban organic wastes. *Biol. Fertil. Soils* **1997**, *24*, 429–434. [CrossRef]
24. Hailegnaw, N.S.; Mercl, F.; Pračke, K.; Száková, J.; Tlustoš, P. High temperature-produced biochar can be efficient in nitrate loss prevention and carbon sequestration. *Geoderma* **2019**, *338*, 48–55. [CrossRef]
25. Mukherjee, A.; Zimmerman, A.R. Organic carbon and nutrient release from a range of laboratory-produced biochars and biochar-soil mixtures. *Geoderma* **2013**, *193*, 122–130. [CrossRef]
26. Lehmann, J.; Rillig, M.C.; Thies, J.; Masiello, C.A.; Hockaday, W.C.; Crowley, D. Biochar effects on soil biota—A review. *Soil Biol. Biochem.* **2011**, *43*, 1812–1836. [CrossRef]
27. Zhou, H.; Zhang, D.; Wang, P.; Liu, X.; Cheng, K.; Li, L.; Zheng, J.; Zhang, X.; Zheng, J.; Crowley, D.; et al. Changes in microbial biomass and the metabolic quotient with biochar addition to agricultural soils: A Meta-analysis. *Agric. Ecosyst. Environ.* **2017**, *239*, 80–89. [CrossRef]
28. Aira, M.; Domínguez, J. Optimizing vermicomposting of animal wastes: Effects of rate of manure application on carbon loss and microbial stabilization. *J. Environ. Manag.* **2008**, *88*, 1525–1529. [CrossRef] [PubMed]
29. Uz, I.; Sonmez, S.; Tavali, I.E.; Çitak, S.; Uras, D.S.; Citak, S. Effect of vermicompost on chemical and biological properties of an alkaline soil with high lime content during celery (*Apium graveolens* L. var. dulce Mill.) production. *Not. Bot. Horti Agrobot.* **2016**, *44*, 280–290. [CrossRef]
30. Liang, C.; Cheng, G.; Wixon, D.L.; Balser, T.C. An Absorbing Markov Chain approach to understanding the microbial role in soil carbon stabilization. *Biogeochemistry* **2011**, *106*, 303–309. [CrossRef]
31. Becagli, M.; Guglielminetti, L.; Cardelli, R. Effects of combined biochar and vermicompost solution on leachate characterization and nitrogen balance from a greenhouse tomato (*Solanum lycopersicum*) cultivation soil. *Commun. Soil Sci. Plant Anal.* **2021**, *52*, 1879–1893. [CrossRef]
32. Gopal, M.; Gupta, A.; Thomas, G. Vermicompost and vermiwash add beneficial micro flora that enhance soil quality and sustain crop growth. *Int. J. Innov. Hortic.* **2012**, *1*, 93–100.
33. Arancon, N.Q.; Edwards, C.A.; Lee, S.; Byrne, R. Effects of humic acids from vermicomposts on plant growth. *Eur. J. Soil Biol.* **2006**, *42*, S65–S69. [CrossRef]
34. Dutta, T.; Kwon, E.; Bhattacharya, S.S.; Jeon, B.H.; Deep, A.; Uchimiya, M.; Kim, K.H. Polycyclic aromatic hydrocarbons and volatile organic compounds in biochar and biochar-amended soil: A review. *GCB Bioenergy* **2017**, *9*, 990–1004. [CrossRef]
35. Swaine, M.; Obrike, R.; Clark, J.M.; Shaw, L.J. Biochar alteration of the sorption of substrates and products in soil enzyme assays. *Appl. Environ. Soil Sci.* **2013**, *2013*, 968682. [CrossRef]
36. Luo, L.; Gu, J.D. Alteration of extracellular enzyme activity and microbial abundance by biochar addition: Implication for carbon sequestration in subtropical mangrove sediment. *J. Environ. Manag.* **2016**, *182*, 29–36. [CrossRef] [PubMed]
37. Lu, H.; Bian, R.; Xia, X.; Cheng, K.; Liu, X.; Liu, Y.; Wang, P.; Li, Z.; Zheng, J.; Zhang, X.; et al. Legacy of soil health improvement with carbon increase following one time amendment of biochar in a paddy soi—A rice farm trial. *Geoderma* **2020**, *376*, 114567. [CrossRef]
38. Günal, E.; Erdem, H.; Demirbaş, A. Effects of three biochar types on activity of β-glucosidase enzyme in two agricultural soils of different textures. *Arch. Agron. Soil Sci.* **2018**, *64*, 1963–1974. [CrossRef]
39. Lim, S.L.; Wu, T.Y.; Lim, P.N.; Shak, K.P.Y. The use of vermicompost in organic farming: Overview, effects on soil and economics. *J. Sci. Food Agric.* **2015**, *95*, 1143–1156. [CrossRef]

40. Wang, P.; Marsh, E.L.; Ainsworth, E.A.; Leakey, A.D.B.; Sheflin, A.M.; Schachtman, D.P. Shifts in microbial communities in soil, rhizosphere and roots of two major crop systems under elevated CO_2 and O_3. *Sci. Rep.* **2017**, *7*, 15019. [CrossRef]
41. Meyer, A.H.; Wooldridge, J.; Dames, J.F. Relationship between soil alteration index three (AI3), soil organic matter and tree performance in a Cripps Pink/M7 apple orchard. *S. Afr. J. Plant Soil* **2014**, *31*, 173–175. [CrossRef]
42. Ghosh, A.; Singh, A.B.; Kumar, R.V.; Manna, M.C.; Bhattacharyya, R.; Rahman, M.M.; Sharma, P.; Rajput, P.S.; Misra, S. Soil enzymes and microbial elemental stoichiometry as bio-indicators of soil quality in diverse cropping systems and nutrient management practices of Indian Vertisols. *Appl. Soil Ecol.* **2020**, *145*, 103304. [CrossRef]
43. Van Huyssteen, I.; Mulidzi, A.R.; Meyer, A.H.; Wooldridge, J. Alteration Index Three facilitates interpretation of β-glucosidase, acid-phosphatase and urease activities in soils irrigated with diluted winery wastewater. *S. Afr. J. Enol. Vitic.* **2020**, *41*, 238–244. [CrossRef]

agriculture

MDPI

Article

Valorization of Quality of Vermicomposts and Composts Using Various Parameters

Monika Jakubus * and Weronika Michalak-Oparowska

Department of Soil Science and Microbiology, Poznan University of Life Sciences, Szydłowska 50, 60-656 Poznan, Poland; ww.michalak@wp.pl
* Correspondence: monika.jakubus@up.poznan.pl

Abstract: Due to the increasing biomass of biowaste it is necessary to manage it rationally. This work presents comparisons and valorization of vermicomposts (VCs) and composts (Cs) prepared from various biowastes generated in households and private gardens, in terms of their practical use. The tested VCs and Cs were subjected to chemical analyses to assess the amounts of macro- (N, P, K, S, Mg, Ca, Na) and micronutrients (Fe, Zn, Mn, Cu, Ni), as well as contents of organic matter (OM), total organic carbon (TOC), humic compounds (HS) and labile and water extractable organic carbon (LC, WEOC). Moreover, humification indexes (HR, HI, DP) were determined. The amounts of macro- and micronutrients, OM, TOC, LC, WEOC were greater for vermicomposts. Regardless of these differences, both vermicomposts and composts were characterized by considerable amounts of organic matter ranging from 325 to 631 $g \cdot kg^{-1}$ and TOC amounting from 82 to 270 $g \cdot kg^{-1}$. Moreover, the tested organic fertilizers were characterized by high contents of N (7–21.5 $g \cdot kg^{-1}$), K (3.7–24.4 $g \cdot kg^{-1}$), Ca (12.2–44.0 $g \cdot kg^{-1}$), Fe (133.1–333.8 $mg \cdot kg^{-1}$) and Mn (71.5–113.8 $mg \cdot kg^{-1}$). The analyzed VCs and Cs did not exceed the permissible amounts of heavy metals (Cr, Pb) and contained a comparable amount and quality of humus compounds. The level of C_{HS} ranged from 29.6 to 41 $g \cdot kg^{-1}$ for vermicomposts, and from 19.8 to 51.8 $g \cdot kg^{-1}$ for composts. The humification indexes indicate that VCs and Cs were well–matured despite different composting conditions. The HI values for VCs ranged from 8.3% to 10% and for Cs amounted from 12.2% to 16.8%. Similarly, the HR values were higher for composts (24.3–33%) in comparison to VCs (15.2–20.1%). Vermicomposting and composting of biowaste is economically and environmentally justified. Fertilizers obtained in the composting process are a valuable source of organic material and nutrients essential for plants and can be safely used in private gardens.

Keywords: agriculture; biowastes; heavy metals; humic compounds; humification indexes; macro- and micronutrients

1. Introduction

Biowastes are a group of municipal wastes, whose mass has recently increased significantly. The prevailing pandemic situation following the COVID-19 outbreak favors such a trend, which was confirmed by numerous authors [1–3]. According to the cited authors, the mass of generated biowastes has increased by 20%, while even a 1.5-fold increase in the mass was noted in some areas. The reasons include primarily the increased consumption and the shopping panic accompanying the imposed lockdown and the resulting fear of problems with the supply of food products. First of all, some products were excessively accumulated, often being perishable food products with a short shelf life, which have not been fully used and constituted a source of food waste, followed by an additional mass of biowaste [2]. Jribi et al. [4] reported that this type of biowaste comprises mainly vegetables, fruit and cereal products, which were either inappropriately stored or were inadequately prepared and had to be disposed of. This mass differing in chemical composition, and hence value, constitutes a valuable material for composting.

Stenmarck et al. [5] reported that 8% of the world's food waste production is destined for home composting, which is about 7.4 kg per capita, per year. However, the group of biowastes also includes biodegradable waste from the maintenance of green areas in cities and private gardens. According to various sources [5–7], biowaste accounts for 30–50% of the total mass of generated municipal wastes. In most countries biowaste is collected selectively, although it is not a rule [8]. Nevertheless, these wastes need to be properly managed in order to reduce their possible negative impact on the environment. Biowaste composting is a popular and cheap solution, in line with the circular economy concept. This process is carried out commercially by appropriate installations; however, as emphasized by Vazquez and Soto [9], home composting complements biodegradable waste management. Moreover, as shown by the local community survey conducted by Jakubus and Michalak-Oparowska [10], home composting of biowaste is gaining popularity and is generally considered acceptable. Vazquez and Soto [9] indicated that recycling of 50% of generated biodegradable waste in domestic composters decreases waste treatment and transportation costs from 34% up to 50%, while simultaneously reducing greenhouse gas emissions by 40% compared to standard landfilling.

Currently, vermitechnology is gaining interest as a method of sustainable and proper biowaste utilization [11–14]. Singh et al. [13] stated that vermicomposting is a technique of biowaste management with the effective support of earthworms. Numerous authors [13,15,16] emphasize the many advantages of vermicompost and its use for agriculture and horticulture. Vermicomposts improve soil health, microbial activity, limit diseases caused by soil-borne pathogenic organisms, as well as stimulate plant growth by changing the physical, chemical and biological properties of the soil. Similar advantages come from the use of traditionally prepared composts [17,18]. However, differences in the quality of the prepared vermicomposts and composts should be considered. This could result from the different biowastes used and the different conditions of the applied process. The standard assessment of organic fertilizer quality is based on the amount of organic matter and N, P and K contents, in view of the fertilizing function of these substances. However, apart from these parameters, it is also reasonable to assess the amount of micronutrients, the content of heavy metals, as well as the quantity and quality of humus compounds. Since organic fertilizers such as vermicomposts and composts are characterized by a considerable amount of organic matter and carbon is present in various combinations, it is important to analyze this aspect in detail. The knowledge on the stability of the carbon compounds in organic matter of vermicomposts and composts is significant from the point of view of their influence on soil fertility [17]. In relation to this, the special role of fulvic and humic acids, as well as labile and water extractable carbon, must be indicated due to their different susceptibilities to solubility and biochemical and microbiological transformations [19,20]. The literature on the evaluation of vermicomposts or composts ignores this aspect, instead focusing on the general characteristics of these organic fertilizers.

In view of the above, the purpose of this study was to present a detailed assessment of vermicomposts and composts in terms of their abundance in macro- and micronutrients and the quantity and quality of humic compounds. Additionally, the amounts of selected heavy metals, as well as easily mineralizable organic carbon forms, were evaluated. Obtained results will facilitate comprehensive valorization of vermicomposts and composts in terms of their quality and compare their practical usability.

2. Materials and Methods

2.1. Composting Procedure and Raw Materials

The aim of the research was achieved on the basis of six different organic material samples (vermicomposts and composts) prepared from selectively collected biowastes. The authors of the study had neither access to raw materials nor influence on the quantity, quality and frequency of deposited wastes for vermicomposting and composting processes. The vermicomposts (VCs) and composts (Cs) were not commercially produced, but only constituted a method of rational management of biomass generated in the household and garden. The following wastes were used for their preparation: food and kitchen wastes from households (VC 1–3); biowastes from the garden: yard trimmings such as plant residues and mowed grass clippings (C4, 6); mixed food and kitchen wastes from households and biowastes from the garden (C5). According to the list of waste referred to Article 7 of Directive 2008/98/EC [21], used wastes belong to the same group of municipal wastes, code 20.

Vermicomposts 1, 2 and 3 were prepared by the vermicomposting process in vermicomposters (Vermitutt Worm Bin). Due to the fact that there are no official recommendations as to the preferred composter and there is a large range of these devices on sale, the operation of the one given in this paper was approximated. The vermicomposter is divided into four partitions (boxes), stacked one above another, with a volume of 15 L each. This design allows for a continuous addition of biowaste and gradual removal of the compost without the need of mixing. A mixture of apple pomace with earthworms (*Eisenia fetida)* was used as an initial input material for the vermicomposting process. The biowaste for the vermicomposter was delivered with a varying frequency and in different amounts, which depended on the activity of the household. Individual boxes were filled within six months. The leachate of composted biowastes was collected in the lowest part of vermicomposter and subsequently it was discharged via a drain valve. The vermicomposting process was carried out at room temperature (±23 °C) and with constant moisture of the mass (±60%).

Composts 4, 5 and 6 were prepared by the aerobic method as a fertilizer for their home gardens by private homeowners. The composting process was carried out in home composters made of thermoplastic. The temperature of the composting process depended on the weather conditions, while the moisture of the composted mass was kept at a similar level (±60%). The organic material (bigger particles were chopped into smaller ones, maximum size of 15–40 mm) was successively collected in containers without any mixing of the bulk volume. This contributed to lesser oxygenation of the mixture inside the composter compared to its top layer. Under such conditions the organic waste mixture was kept for a year. After this time, the whole mass was mixed to homogenize it and then transferred to dark plastic bags to complete the maturation stage.

The vermicompost samples were collected from individual boxes of the vermicomposter and the bulk samples represented approximately 80–100% of the total box volume. The compost samples were collected from the bags after their contents had been mixed. The samples of organic materials were dried at 105 °C for a period of 12 h. The dried samples were ground into a fine powder and stored in plastic bags at a temperature of 4 °C.

2.2. Chemical Analysis of the Compost

The chemical analyses were conducted on dried samples. The reaction (pH) and electrolytic conductivity (EC) of the tested materials were determined in an aqueous solution at a ratio of 1:10. Total organic carbon (TOC), nitrogen (N) and sulfur (S) contents in VCs and Cs were assayed using a Vario Max CNS elemental analyzer. On the basis of TOC and N total amounts the C:N ratio was calculated following equation: $C{:}N = \frac{TOC}{Ntot}$. The loss-on-ignition method was used to determine organic matter (OM) of vermicomposts and composts. For this purpose samples were subjected to dry combustion for 6 h at a temperature of 550 °C. The ash of VCs and Cs after combustion was used to determine the total amounts of macro- and micronutrients as well as heavy metals. Thus the ash was dissolved in 5 mL of 6 mol·dm^3 HCl and diluted to a constant volume with distilled water. In the obtained extracts K, Ca, Mg and Na assessment was performed using atomic absorption spectrophotometry, while total phosphorus (P) content was measured colorimetrically by the vanadium–molybdenum method [22]. The determinations of microelements and heavy metals were performed in the same solutions that were used to determine macronutrients. The micronutrients (Fe, Mn, Zn, Cu, Ni) and heavy metals (Pb, Cr) were evaluated using atomic absorption spectrophotometry with a Varian Spectra AA 220 FS apparatus.

Humus fractionation of VCs and Cs was performed according to the method proposed by Kononova and Bielczikova, in which humic substances (HS) were determined in a mixture of 0.1 mol·dm^{-3} $Na_4P_2O_7$ + 0.1 mol·dm^{-3} NaOH solution [23]. The fulvic acid fraction (FA) was separated after precipitation of humic acids at pH 1.5 (HA). Carbon in the obtained fractions (C_{HS} and C_{FA}) was oxidized by 0.1 mol·dm^{-3} $KMnO_4$ in the H_2SO_4 medium. Humic acid carbon (C_{HA}) was calculated by subtracting C_{FA} from C_{HS}. Optical density ($Q_{4/6}$) of the humic substances was determined at 465 nm and 665 nm. Additionally, the samples were used to analyze labile and water extractable organic carbon. The labile carbon (LC) was assessed by $KMnO_4$ oxidation [24], while water extractable organic carbon (WEOC) was determined according to the method presented by Ghani et al. [25] with the final determination of organic carbon by wet combustion [26].

Additionally, three popular indexes, i.e., the humification ratio (HR), humification index (HI) and the degree of polymerisation (DP), were used in this study. The humification indexes were calculated using the following equations [27]:

$$\mathrm{HR}\ (\%) = \frac{CHS}{TOC} \cdot 100 \tag{1}$$

$$\mathrm{HI}\ (\%) = \frac{CHA}{TOC} \cdot 100 \tag{2}$$

$$\mathrm{DP} = \frac{CHA}{CFA} \tag{3}$$

2.3. Statistical Analysis

The data presented in the paper are means of three replications. The data were compiled applying one–way ANOVA. Each of the sixteen parameters was tested independently using the F-test at the significance level $\alpha = 0.95$. The null hypothesis assumption was that the mean values of the examined parameter are equal for each of the analyzed vermicomposts and composts against the alternative hypothesis that not all the means are equal. As a result of the rejection of the null hypothesis the least significant differences were calculated using the Tukey test at the significance level $\alpha = 0.05$. Tukey's analysis was performed to distinguish homogeneous groups among the analyzed vermicomposts and composts. In addition, Person's correlation coefficients were calculated for the analyzed parameters. Moreover, for pairs (x, y) of correlated parameters estimates of simple regressions of the form can be determined (regression model): $y = \beta_0 + \beta_1 x$, where the regression parameter β_1 shall be interpreted as follows: if parameter x increases by one unit, parameter y increases (decreases) by β_1 units. The data were analyzed using the STATOBL software working in the Windows environment.

3. Results and Discussion

The basic assumption of biowaste composting is the possibility of reusing organic matter and nutrients contained in it. This approach to the production of biomass is consistent with the concept of circular economy and it is the most rational method of biowaste management [8,18,28]. Considering the final use of compost or vermicompost for agricultural or horticultural purposes, their quality is of greater importance, including the abundance of organic matter and the essential nutrients. Additionally, selected physicochemical properties of organic materials, such as reaction and electrolytic conductivity, are also significant. Reaction is a key factor in determining the transformation of organic compounds and the availability of nutrients for plants. Therefore, it is also important to analyze this parameter in the vermicomposts and composts. In the tested VCs 1–3 the pH values ranged from 7.9 to 8.4, while in Cs 4–6 from 6.5 to 7.2 (Table 1). Regardless of the biowaste composting method, the EC values were comparable to the level of 3.1 to 3.7 mS·cm^{-1} (Table 1). Singh et al. [13] gave similar ranges of pH and EC values for various vermicomposts. Moreover, these authors citing Wong et al. [29] reported that VCs with an EC below 4 mS·cm^{-1} is appropriate to be used as fertilizer for plant growth. Sciubba et al. [30]

presented similar values of pH and EC for various composts, while Ibrahim et al. [31] reported higher EC values.

Table 1. The values of pH and EC of analyzed vermicomposts and composts.

Parameter	VC1	VC2	VC3	C4	C5	C6
pH	7.9	8.4	8.3	7.2	6.8	6.5
EC ($mS \cdot cm^{-1}$)	3.5	3.6	3.7	3.3	3.7	3.1

In Poland, VCs and Cs dedicated to agricultural use have to meet specific threshold amounts of N, P, K, organic matter and heavy metals [32]. According to the above-mentioned Regulation, the content of OM must be at least 30% d.m., the amount of potassium (K_2O) and phosphorus (P_2O_5) should be more than 0.2% d.m., while the total N value should be min. 0.3% d.m. The limit value of organic matter at 31.5% for composts was indicated by Vazquez and Soto [9].

Comparing the above-mentioned threshold values of OM with those obtained in this study, only C5 failed to meet the requirement for OM because for this compost a 150 $g \cdot kg^{-1}$ of OM was recorded. For VCs 1–3 the organic matter contents were considerably higher and amounted from 456 to 631 $g \cdot kg^{-1}$, while C4 and C6 contained 325 and 407 $g \cdot kg^{-1}$, respectively (Figure 1). Consequently, the vermicomposts were characterized by higher TOC contents ranging from 150 to 270 $g \cdot kg^{-1}$ than composts 4–6 (82 to 163 $g \cdot kg^{-1}$) (Figure 1). Similar TOC contents for VC were reported by Singh et al. [13] and Ramnarain et al. [33]. For composts, Jakubus [17], Sciubba et al. [30] and Ibrahim et al. [31] found comparable amounts of TOC to those presented in this study. Significantly higher amounts (2- to 3-fold greater) of OM and TOC were indicated by Yu et al. [34] for compost, but in this case the analyzed compost was prepared on the basis of cow manure, which may explain such a difference.

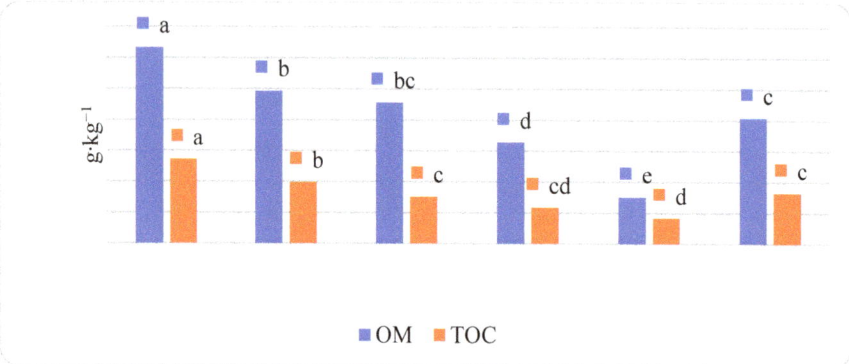

Figure 1. Contents of organic matter (OM) and total organic carbon (TOC) in analyzed vermicomposts and composts (the same lowercase letters indicate homogeneous groups of VCs and Cs).

Compost introduced into the soil first of all provides a significant amount of organic matter; however, it needs to be remembered that not only the amount but also the quality of the applied organic matter is very important. Therefore, apart from the basic determination of the OM and TOC contents, it is necessary to analyze the quantity and quality of humic compounds and easily mineralized carbon compounds. Especially the latter compounds directly affect the rate and direction of changes of native and introduced carbon in the soil, creating soil fertility. The changes of water extractable organic compounds can reflect the transformation degree of organic matter and the stability of materials during the composting process [34]. The authors [24,29,34,35] emphasized the importance of WEOC transformations. They considered WEOC as a component of the labile and the most active fraction of organic waste. It is a sensitive measure of subtle changes in organic matter, and it could directly reflect the organic matter transformation process. Consequently, the composition of WEOC was suggested as a better indicator of stability for the OM. In view of the above, the analyzed VCs and Cs were characterized in the terms of labile carbon and water extracted organic carbon. Again, vermicomposts were characterized by a higher content of easily mineralized carbon compounds (LC and WEOC) than composts. As shown by the data in Figure 2, the amounts of LC ranged from 0.94 to 1.02 $g \cdot kg^{-1}$ and WEOC from 6.08 to 7.32 $g \cdot kg^{-1}$ for VC 1–3, while for Cs 4–6 it was from 0.23 to 0.58 $g \cdot kg^{-1}$ LC and from 1.9 to 4.43 $g \cdot kg^{-1}$ for WEOC. On this basis it can be assumed that after introducing VCs into the soil, they will accelerate the mineralization processes, becoming a source of both easily activated nutrients for plants and a source of energy for microorganisms.

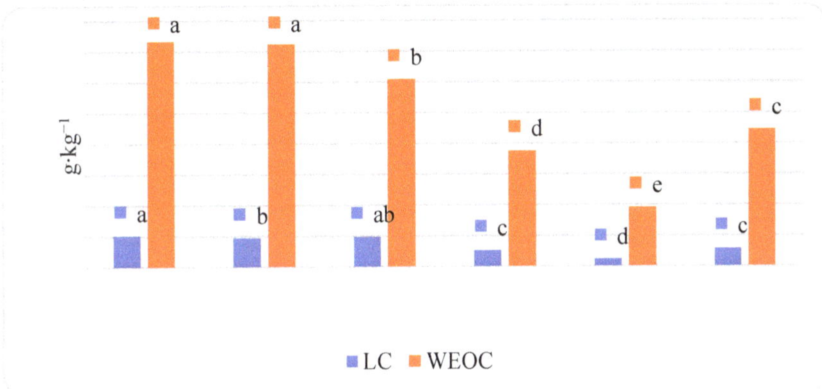

Figure 2. Contents of labile carbon (LC) and water extractable organic carbon (WEOC) in analyzed vermicomposts and composts (the same lowercase letters indicate homogeneous groups of VCs and Cs).

A detailed quality analysis of VCs and Cs should also include the humic compounds strongly determining chemical properties of organic fertilizers. Humic substances are the result of the humification process of organic compounds, which may proceed at different rates depending on the environmental conditions. Humic substances mainly consist of fulvic and humic acids, wherein HAs have a more complex structure than FAs. Fulvic acids are compounds weakly polymerized and relatively readily undergoing chemical and microbiological changes, which results in their considerable solubility and mobility. In turn, HAs are generally recognized as being non-degradable or sparsely degradable compounds with a strongly polymerized structure [34,36]. Generally immature composts contain a high FA content and a relatively low HA content, while HA dominates in mature composts [36].

In the present study, the vermicomposting and composting process differed significantly in terms of the prevailing conditions (outdoor, indoor, addition of earthworms vs. no such addition, various biowastes, duration of the process). However, regardless of the above, the tested materials had comparable quantitative levels of humic substances. The amount of C_{HS} ranged from 29.6 to 41 $g \cdot kg^{-1}$ for the vermicomposts, and from 19.8 to 51.8 $g \cdot kg^{-1}$ for the composts (Figure 3). The amounts of C_{FA} and C_{HA} for the individual organic materials were comparable, without being significantly different. The C_{FAs} amounts of VCs 1–3 ranged from 15.2 to 19.8 $g \cdot kg^{-1}$, and for Cs 4–6 it was from 9.9 to 26.6 $g \cdot kg^{-1}$. On the other hand, the C_{HAs} ranged from 14.4 to 22.4 $g \cdot kg^{-1}$ and from 9.9 to 25.2 $g \cdot kg^{-1}$ for VCs 1–3 and Cs 4–6, respectively (Figure 3).

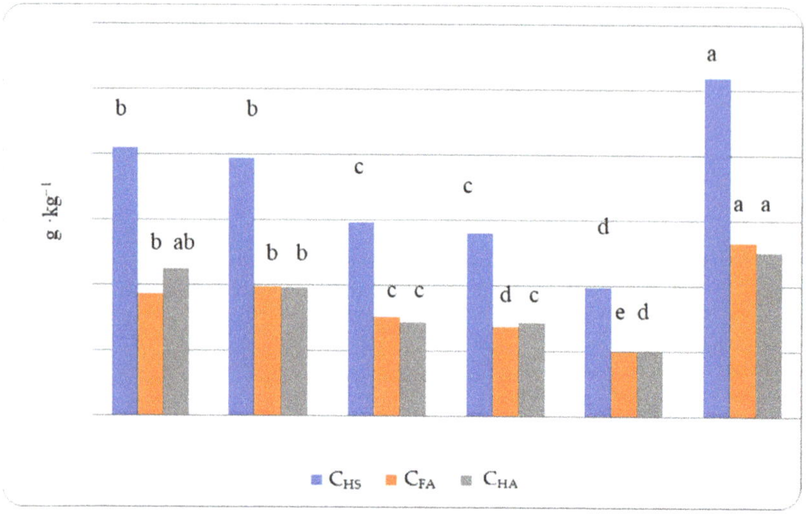

Figure 3. Contents of humic substance (C_{HS}), fulvic acids (C_{FA}) and humic acids (C_{HA}) in analyzed vermicomposts and composts (the same lowercase letters indicate homogeneous groups of VCs and Cs).

When assessing the quality of the vermicomposts and composts the evaluation of their stability and maturity is also essential. This is an extremely important element of composting because an unstable and immature organic material may have adverse effects on plant growth and the environment. First of all, the presence of volatile chemicals, such as organic acids toxic for plants, must be underlined. Additionally, an incompletely matured compost with a higher C:N ratio can lead to a biological block of nitrogen, also known as "nitrogen starvation" [37]. There is also disagreement among various authors in relation to the specific limit value of C:N. According to Chen et al. [38], Gomez-Brandon et al. [39] and Singh et al. [40] the C:N ratio for the finished compost should range from 10:1 to 15:1. Similar conclusions in their own research were shown by Vazquez and Soto [9], who proved that C:N for mature composts should be 9–16:1. However, Antil et al. [41] stated that the C:N ratio in composts should range from 15:1 up to 20:1. Additionally, Asquer et al. [42] indicated that the C:N ratio needs to be lower than 20:1. Taking into consideration a broad range of the C:N ratio at 9–20:1 as the criterion, only the analyzed VC3 failed to meet this threshold (Table 2). Jakubus [17] and Sciubba et al. [30] indicated similar values of the C:N ratio in composts, while Ibrahim et al. [31] showed a higher value (18:1) of the discussed parameter. For vermicomposts, Balachandar et al. [43] indicated higher values of C:N (16.56–17.55) than those in the presented study.

Table 2. The C:N ratio and humification indexes obtained for analyzed vermicomposts and composts.

Compost	HI (%)	HR (%)	DP (C_{HA}:C_{FA} Ratio)	$Q_{4/6}$	C:N
VC1	8.3	15.2	1.2	8.1	17:1
VC2	10.0	20.1	1.0	8.1	9:1
VC3	9.6	19.7	1.0	7.7	8:1
C4	16.8	33.0	1.0	6.6	9:1
C5	12.2	24.3	1.0	7.2	12:1
C6	15.5	31.8	1.0	7.1	9:1

Vermicomposts and composts produced from similar raw organic materials should be assessed by various humification indexes, especially when their maturity is considered. According to Li et al. [36], the humified fraction of OM is the most important and responsible for fertility functions. Thus, evaluation of the humification degree is an agronomic criterion for compost quality. In practice, HR, HI and DP are generally accepted and helpful in compost maturity evaluation. Table 2 contains values of the other parameters enabling the assessment of VCs and Cs in terms of their maturity. It is assumed that mature composts should have an HI value above 30% [44]. For such a criterion being adopted here, none of the analyzed organic fertilizers reached this value because, as indicated in Table 2, HI values ranged from 8.3% (VC1) to 16.8% (C4). For vermicompost No. 1 and compost No. 4 the lowest (15.2%) and the highest (33.0%) HR values were determined (Table 2). However, there is no limit HR value specified in the literature, which could be helpful in assessing compost maturity. Nevertheless, results presented by various authors showed similar values of HR and HI for composts [17,42,45] and they are comparable with those given in this study. The polymerization degree (PD) expressed as the C_{HA}:C_{FA} ratio is widely used to describe the relative speed of the HA and FA transformation, as well as the maturity of the compost. Azim et al. [37], on the basis of a literature review, stated that the correct threshold value of PD needs to be greater than one and simultaneously, according to Alavarenga et al. [46], not higher than 2.5. Taking into account this threshold it may be assumed that all the analyzed VCs and Cs were well-matured because the values obtained for them ranged from 1.0 to 1.2 (Table 2).

The $Q_{4/6}$ ratio is negatively related to the aromatic polycondensation degree and molecular weight of humic substances. High $Q_{4/6}$ values imply the presence of low molecular weight aromatic molecules, which in contrast to $Q_{4/6}$ low values indicate high contents of large molecular weight molecules, such as humic-like compounds, usually present in well-matured organic materials [46]. In the present study, the analyzed vermicomposts and composts also showed comparable values of optical density expressed as $Q_{4/6}$, which range from 6.6 (C4) to 8.1 (VC1 and 2) (Table 2). Similar values of the discussed parameter were showed by Lv et al. [19]. In turn, Ozdemir et al. [28] reported the value range of optical density from 3.23 to 8.8 to be adequate for compost maturation. Based on this statement, the VCs and Cs analyzed in this study were well maturated.

When assessing the macronutrient abundance of the vermicomposts and composts it should be emphasized that their amounts in the vermicomposts are significantly greater than in the composts. Moreover, the amounts of macronutrients generally did not differ significantly between the VCs and Cs (Table 3). As previously mentioned, according to Polish law, VCs and Cs must meet specific requirements for NPK content. As results from the data in Table 3, the amounts of N ranged from 7 to 17.6 g·kg^{-1} for C4–6 and from 16.1 to 21.5 g·kg^{-1} for VCs, which significantly exceeds the minimum limit specified in the above-mentioned documents. Similar N amounts in vermicomposts were indicated by Singh et al. [13] and Ramnarain et al. [33]. Additionally, a similar range of N values in composts can be found in the literature data [17,31,37]. Higher N amounts were shown by Sciubba et al. [30] and Yu et al. [34].

Table 3. Macronutrient total amounts in analyzed vermicomposts and composts (g·kg^{-1}).

Macronutrient	VC1	VC2	VC3	C4	C5	C6
N	16.1 c	21.5 a	18.4 b	10.0 d	7.00 e	17.6 bc
P	4.1 c	6.0 a	5.8 a	1.2 d	5.9 b	4.8 b
K	18.0 b	24.4 a	17.2 b	4.5 c	3.7 c	5.5 c
S	4.60 a	4.6 a	3.5 b	2.7 c	1.4 d	2.9 c
Ca	20.6 c	34.1 b	44.0 a	13.4 d	12.2 d	12.3 d
Mg	4.1 c	8.7 b	13.5 a	1.7 d	1.4 d	1.6 d
Na	1.6 b	2.1 a	1.5 b	0.9 c	1.4 b	0.9 c

The same lowercase letters indicate homogeneous groups of VCs and Cs.

In comparison to the Regulation of the Minister of Agriculture and Rural Development [32], the obtained amounts of P and K (Table 3) were considerably higher, especially for VCs. The contents of P and K in VCs 1–3 ranged from 4.1 to 5.8 g·kg^{-1} and from

17.2 to 24.4 g·kg^{-1}, respectively. Cs 4–6 presented lower amounts amounting from 1.2 up to 5.9 g·kg^{-1} of P and from 3.7 to 5.5 g·kg^{-1} for K. Singh et al. [13] for vermicomposts, which showed significantly higher (4–5 times) amounts of P and comparable K. On the other hand, the composts tested by Jakubus [17] had a lower quantitative level of P and a higher levels of K.

Considerably higher contents of P and K in composts were reported by Sciubba et al. [30], Ibrahim et al. [31] and Yu et al. [34]. For plant development, other macronutrients such as S, Ca, Mg and Na are also important, but they are not subject to routine and mandatory verification and are often ignored in scientific research when valorizing VCs and Cs. Jakubus [17], when analyzing the fertilization quality of various composts, found a similar level of Mg and definitely lower contents of Ca, Na and S compared to the data obtained in this study. Particular attention in this context needs to be paid to 8-fold higher amounts of sulfur given by the cited author (Table 3).

Vermicomposts and composts, apart from macronutrients, are also rich in micronutrients. Micronutrients are taken up in smaller amounts than the previously characterized macronutrients; however, they play important functions in most physiological and biochemical processes [47]. Generally, vermicomposts contained higher amounts of Ni, Cu and Fe and lower levels of Zn and Mn in relation to the contents specified in the composts (Table 4). Regardless of the above, the amounts of micronutrients found in the vermicomposts in this study were significantly lower than those reported by Ramnarain et al. [33]. Compared to the values given in this study, Jakubus [17], Ibrahim et al. [31] and Rodrigues et al. [48] indicated greater amounts of micronutrients in the composts prepared on the basis of various biowastes. When valorizing the quality of vermicomposts or composts, we must take into account the fact that they may be loaded with heavy metals. The group of heavy metals includes many elements, which are both micronutrients necessary for plants, such as Cu, Zn, Mn, Ni, and toxic ones, such as Pb or Cr. Taking into account the negative impact of heavy metals on the soil environment and their easy incorporation into the food chain, vermicomposts and composts must meet the criteria for the content of heavy metals. According to the Regulation of the Minister of Agriculture and Rural Development on 18 June 2008 [32], vermicompost or compost cannot exceed, among other things, 100 mg·kg^{-1} Cr, 60 mg·kg^{-1} Ni and 140 mg·kg^{-1} Pb. The EU guidelines [49] in this regard are more restrictive because the amount of Ni in composts cannot be higher than 20 mg·kg^{-1}, Pb higher than 45 mg·kg^{-1}, and Cr greater than 70 mg·kg^{-1} d.m. Contents of these metals in the analyzed vermicomposts and composts showed that all of them meet the national and European standards, since the amounts of Ni, Pb and Cr were significantly lower (Table 4, Figure 4) in relation to the above-mentioned threshold values. The studies of Balachandar et al. [43] also indicated the amount of heavy metals in vermicomposts to be significantly below the permissible limits. Nevertheless, it should be noted that vermicomposts were generally characterized by higher amounts of Cr and Pb compared to those specified in the composts (Figure 4). For VCs 1–3, the Cr contents ranged from 8.8 to 19.2 mg·kg^{-1} and for Pb from 4.9 to 12.1 mg·kg^{-1}. On the other hand, for Cs 4–6, the Cr amounts were significantly lower and ranged from 3.4 to 6.5 mg·kg^{-1}. Lead levels in composts vary significantly from 4.0 to 20.5 mg·kg^{-1} (Figure 4). Jakubus [17], Ibrahim et al. [31] and Rodrigues et al. [48] gave significantly higher amounts of Cr, Pb and Ni for composts.

Table 4. Micronutrient total amounts in analyzed vermicomposts and composts (mg·kg^{-1}).

Micronutrient	VC1	VC2	VC3	C4	C5	C6
Cu	13.3 bc	18.4 a	16.4 ab	9.4 cd	9.6 cd	8.1 d
Zn	3.0 d	5.4 b	4.3 c	8.8 a	6.1 b	4.3 c
Mn	71.5 c	102.5 ab	85.1 bc	93.1 b	113.8 a	99.5 ab
Ni	4.0 bc	6.4 a	5.3 ab	5.8 ab	3.0 c	2.9 c
Fe	188.5 c	333.8 a	273.8 b	214.3 c	198.6 c	133.1 d

The same lowercase letters indicate homogeneous groups of VCs and Cs.

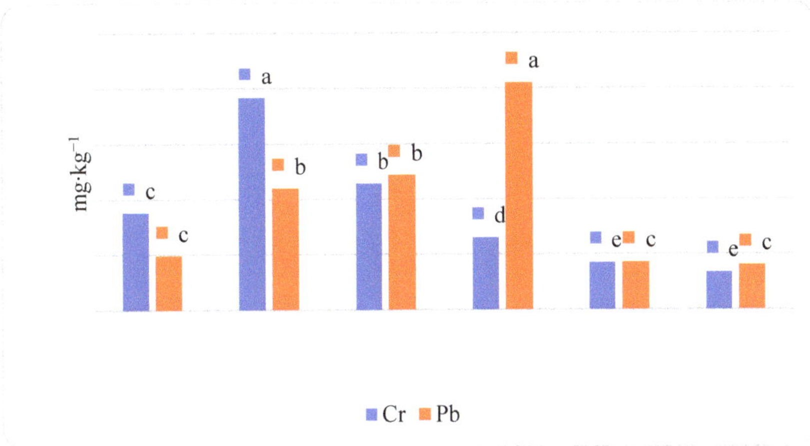

Figure 4. Heavy metal total contents (Cr, Pb) in analyzed vermicomposts and composts (the same lowercase letters indicate homogeneous groups of VCs and Cs).

The chemical composition of organic additives is important especially when they are used as fertilizers. In this case the potential transformations of organic compounds (organic matter, carbon, nitrogen, sulfur and phosphorus compounds) introduced into the soil with composts or vermicomposts are also important. Apart from the value of particular parameters, it is very interesting to see how these parameters interact with each other. Such mutual relationships can be evaluated based on the correlations between the analyzed parameters, an aspect which was also assessed in this study (see Supplementary Material, Tables S1 and S2). However, their correlations may be the same and fail to explain the interrelationships or state whether if the value of one of them increases, the other increases or decreases proportionally. Such information can be obtained from the linear regression model.

Regardless of the analyzed vermicomposts or composts, a strong influence of the organic matter and WEOC on other parameters was noted. In vermicomposts, the amount of OM positively influenced the content of TOC, C_{SH}, LC and WEOC, which was particularly noticeable in the case of TOC and C_{SH}. Together with an increase in OM by 1 g·kg^{-1} the average amount of TOC or C_{SH} can increase by 0.63 and 0.05 g·kg^{-1} in vermicomposts, respectively (Figure 5). The increasing content of OM at the same time can strongly decrease the amount of P, Mg and Ca. The increment of OM by 1 g·kg^{-1} can cause a reduction in the content of the above-mentioned nutrients by 0.12 (Ca), 0.05 (Mg) and by 0.02 (P) g·kg^{-1} (Figure 6). The negative although slightly weaker interactions were noted between the amount of OM and metal contents: Zn, Ni, Cu, Fe and Pb (Table S1). In vermicomposts, an effect of WEOC content on the amounts of TOC, C_{SH} and S was also found. According to the linear regression presented in Figure 7, at an WEOC increase by 1 g·kg^{-1} TOC increased by 67.61 g·kg^{-1}, C_{SH} by 8.02 g·kg^{-1} and S by 0.84 g·kg^{-1}, respectively.

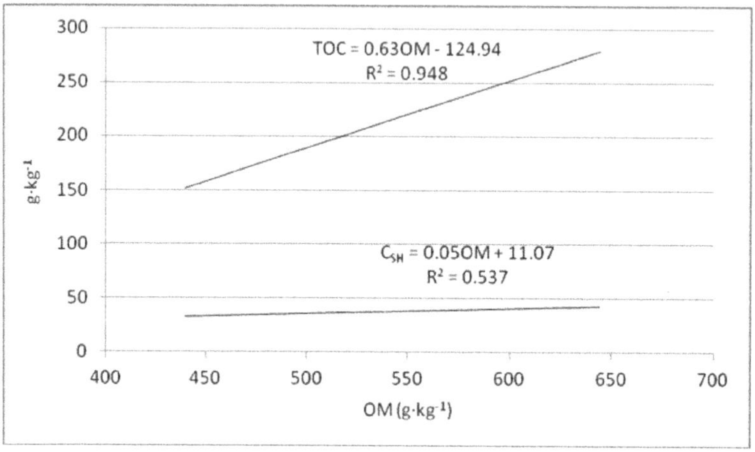

Figure 5. Linear regression for organic matter (OM) and humic substance (C_{SH}) and total organic carbon (TOC) in vermicomposts.

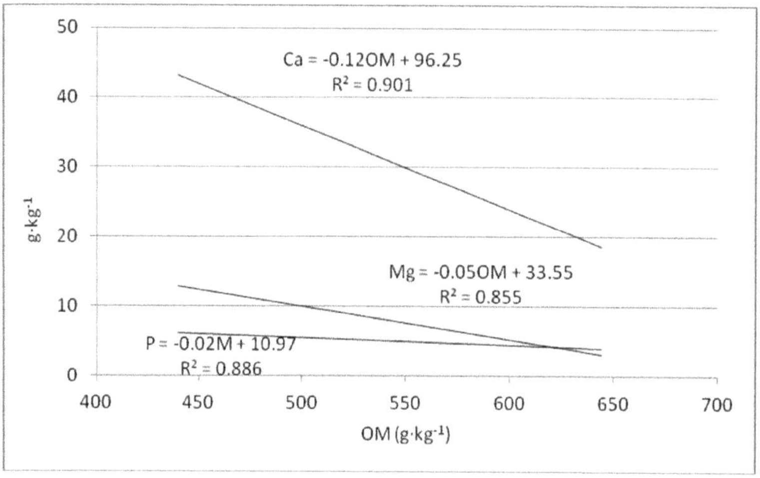

Figure 6. Linear regression for organic matter (OM) and Ca, P and Mg in vermicomposts.

On the other hand, in composts, the gain of OM by 1 g·kg^{-1} was accompanied with an average TOC increment by 0.33 g·kg^{-1}, C_{SH} by 0.12 g·kg^{-1} and the N by 0.04 g·kg^{-1} (Figure 8). The effect of OM on the amount of WEOC, LC or S was positive, although less marked (Table S2). The amounts of WEOC also influenced selected compost parameters and their dependences were directly proportional. An increase in WEOC by 1 g·kg^{-1} caused an increase in OM by 86.52 g·kg^{-1}, TOC by 25.0 g·kg^{-1} and C_{SH} by 10.79 g·kg^{-1}, respectively (Figure 9). Moreover, together with an increment of WEOC by 1 g·kg^{-1}, the gain in the amount of LC by 0.142 g·kg^{-1}, S by 0.65 g·kg^{-1} and N by 3.55 g·kg^{-1} was observed (Figure 10).

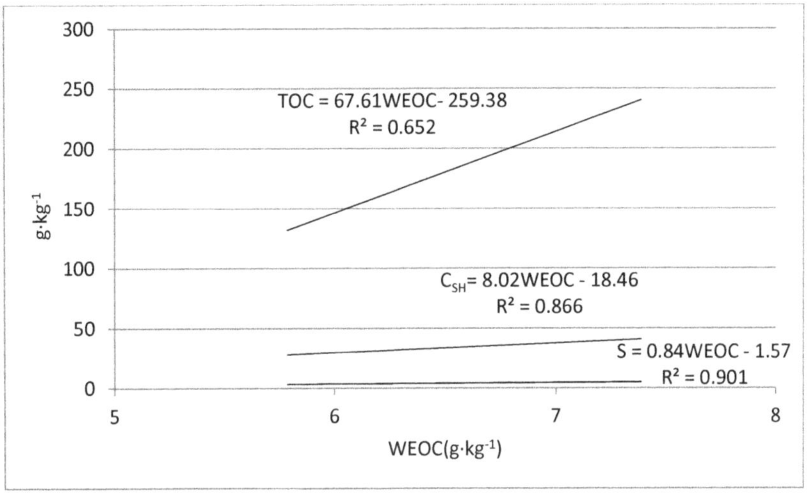

Figure 7. Linear regression for water extractable organic carbon (WEOC) and humic substance (C_{SH}), total organic carbon (TOC) and S in vermicomposts.

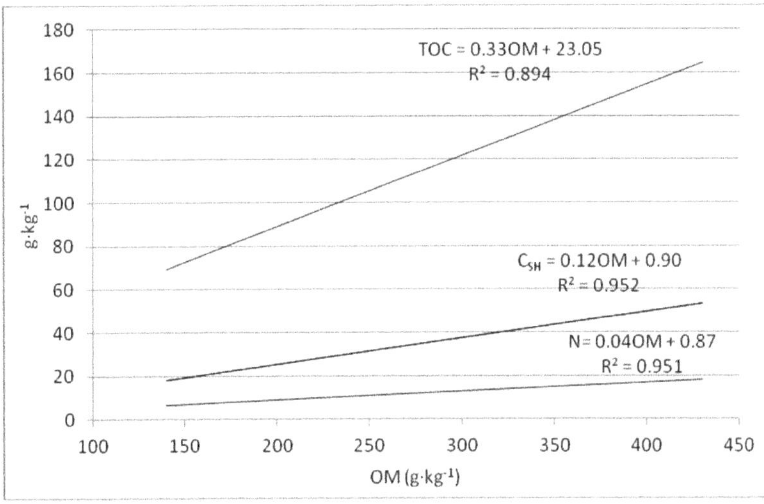

Figure 8. Linear regression for organic matter (OM) and humic substances (C_{SH}), N and total organic carbon (TOC) in composts.

The relationships shown above emphasize the essential importance of both organic matter and water extractable organic carbon in shaping the quality of the vermicomposts and composts. A strong influence was observed with regard to total organic carbon, humic substance, labile carbon, as well as the nutrients integrally bound to the organic matter of vermicomposts and composts, i.e., nitrogen and sulfur. Taking into account the fact that the above-mentioned C compounds are more or less susceptible to decomposition, it can theoretically be assumed that they may play an important role in the transformation of VCs and Cs in the soil. Obviously, to verify the above assumption experimental studies need to be conducted. On the one hand, the rapid degradation of easily mineralizable C compounds (LC, WEOC) can contribute to enhancing the microbial activity of the soil as well as releasing N and S from easily mineralizable combinations. On the other hand, the introduction of organic matter from VC or C to the soil will increase the amount of humic substances—persistent C compounds that determine the improvement of sorption and buffer the properties of the soil.

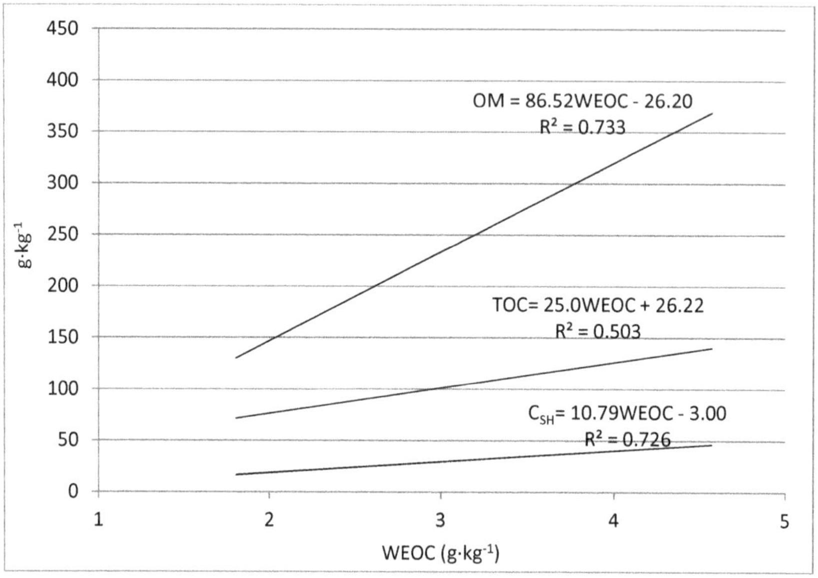

Figure 9. Linear regression for water extractable organic carbon (WEOC) and humic substance (C_{SH}), total organic carbon (TOC) and organic matter (OM) in composts.

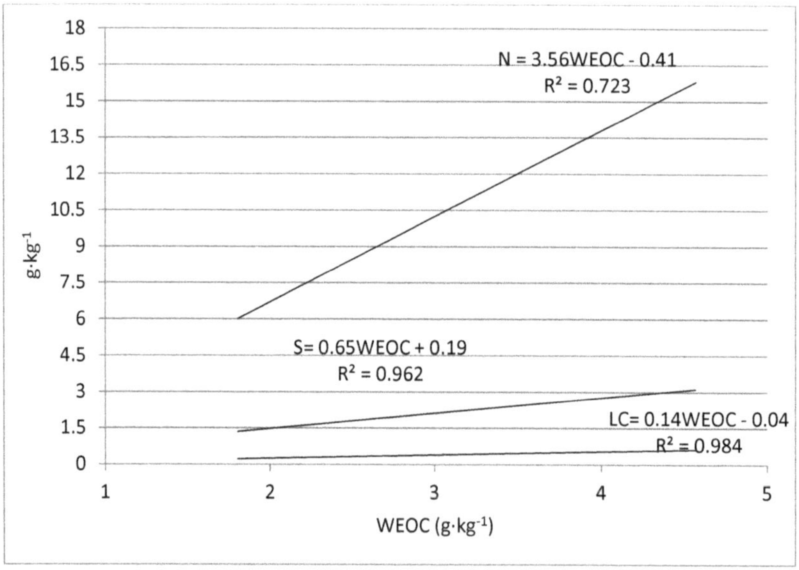

Figure 10. Linear regression for water extractable organic carbon (WEOC) and labile carbon (LC), N and S in composts.

4. Conclusions

On the basis of the obtained results it can be concluded that both the vermicomposts and composts were of good quality, serving as a valuable source of organic matter and nutrients for plants and thus they can be used for private gardening purposes. The fact that the content of heavy metals in the VCs and Cs did not exceed the permissible standards, high safety of their soil application should be underlined. Compared to the composts, vermicomposts were more abundant in macro- and micronutrients. For them, higher amounts of organic matter, TOC, LC and WEOC were also recorded. In view of the above, vermicomposts seem to be better fertilizers than traditionally prepared composts. In this context, not only the amount of introduced nutrients essential for plants should be emphasized, but also the load of easily mineralized carbon compounds (LC and WEOC). However, despite the differences resulting from the biowaste used, as well as specificity of the process, the tested vermicomposts and composts did not vary in terms of the quantity and quality of humus compounds described by the DP, $Q_{4/6}$ values or the amount of C_{HA} and C_{FA}. Regardless of the humification indexes (HR, HI, DP) indicating a satisfactory maturity of the tested materials, the high $Q_{4/6}$ values underlined low optical density and poorly polymerized humic compounds in the analyzed organic fertilizers.

Supplementary Materials: The following supporting information can be downloaded at: https://www.mdpi.com/article/10.3390/agriculture12020293/s1, Table S1: Correlation coefficients matrix for vermicomposts; Table S2: Correlation coefficients matrix for composts

Author Contributions: Conceptualization, M.J.; methodology, M.J.; software, M.J. and W.M.-O.; validation, M.J.; formal analysis, M.J.; investigation, M.J. and W.M.-O.; resources, M.J.; data curation, M.J.; writing—original draft preparation, M.J.; writing—review and editing, M.J.; visualization, M.J. and W.M.-O.; supervision, M.J.; project administration, M.J.; funding acquisition, M.J. All authors have read and agreed to the published version of the manuscript.

Funding: The publication was co-financed within the framework of the Ministry of Science and Higher Education Program as "Regional Initiative Excellence" in years 2019–2022, Project No. 005/RID/2018/19.

Institutional Review Board Statement: Not applicable.

Informed Consent Statement: Not applicable.

Data Availability Statement: Not applicable.

Conflicts of Interest: The authors declare no conflict of interest.

References

1. Fan, Y.; Jiang, P.; Hemzal, M.; Klemes, J.J. An update of COVID-19 influence on waste management. *Sci. Total Environ.* **2021**, *754*, 142014. [CrossRef] [PubMed]
2. Makanjuola, O.; Arowosola, T.; Chenyu, D. The Utilization of Food Waste: Challenges and Opportunities. *J. Food Chem. Nanotechnol.* **2020**, *6*, 182–188. [CrossRef]
3. Sarkodie, S.A.; Owusu, P.A. Impact of COVID–19 pandemic on waste management. *Environ. Dev. Sustain.* **2020**, *26*, 7951–7960. [CrossRef] [PubMed]
4. Jribi, S.; Ismail, H.B.; Doggui, D.; Debbabi, H. COVID–19 virus outbreak lockdown: What impacts on household food wastage? *Environ. Dev. Sustain.* **2020**, *22*, 3939–3955. [CrossRef]
5. Stenmarck, A.; Jensen, C.; Quested, T.; Moates, G. Estimates of European Food Waste Levels. Fusion 2016. Available online: https://www.eu-fusions.org/phocadownload/Publications/Estimates%20of%20European%20food%20waste%20levels.pdf (accessed on 15 December 2021).
6. Delgado, M.; Lopez, A.; Cuartas, M.; Rico, C.; Lobo, A. A decision support tool for planning biowaste management systems. *J. Clean Prod.* **2020**, *242*, 118460. [CrossRef]
7. Penteado, C.S.C.; Soare de Castro, M.A. COVID-19 effects on municipal solid waste management: What can effectively be done in the Brazilian scenario? *Resour. Conserv. Recycl.* **2021**, *164*, 105152. [CrossRef]
8. Jakubus, M.; Stejskal, B. Municipal solid waste management systems in Poland and the Czech Republic. A comparative study. *Environ. Prot. Eng.* **2020**, *46*, 61–78. [CrossRef]
9. Vazquez, M.A.; Soto, M. The efficiency of home composting programmes and compost quality. *Waste Manag.* **2017**, *64*, 39–50. [CrossRef]

10. Jakubus, M.; Michalak-Oparowska, W. Social participation in the biowaste disposal system before and during the COVID-19 pandemic. A case study for Poznań. *Environ. Prot. Eng.* **2021**, *47*, 109–123. [CrossRef]
11. Soobhany, N. Insight into the recovery of nutrients from organic solid waste through biochemical conversion processes for fertilizer production: A review. *J. Clean Prod.* **2019**, *241*, 118413. [CrossRef]
12. Biruntha, M.; Karmegam, N.; Archana, J.; Karunau, B.S.K.; Arockia, J.J.P.; Alamuralikrishnan, B.; Chang, S.W.; Ravindran, B. Vermiconversion of biowastes with low-to-high C/N ratio into value added vermicompost. *Bioresour. Technol.* **2020**, *297*, 122398. [CrossRef] [PubMed]
13. Singh, S.; Singh, J.; Kandoria, A.; Quadar, J.; Bhat, S.A.; Chowdhary, A.B.; Vig, A.P. Bioconversion of different organic wastes into fortified vermicompost with the help of earthworms: A comprehensive review. *Int. J. Recycl. Org. Waste Agric.* **2020**, *9*, 432–439. [CrossRef]
14. Zziwa, A.; Jagwe, J.; Kizito, S.; Kabeng, I.; Komaken, A.J.; Kayondo, H. Nutrient recovery from pineapple waste through controlled batch and continuous vermicomposting systems. *J. Environ. Manag.* **2021**, *279*, 111784. [CrossRef] [PubMed]
15. Biswas, I.; Mitra, D.; Senapati, A.; Mitra, D.; Chattaraj, S.; Ali, M.; Basak, G.; Panneerselvam, P.; Das Mohapatra, P.K. Valorization of vermicompost with bacterial fermented chicken feather hydrolysate for the yield improvement of tomato plant: A novel organic combination. *Int. J. Recycl. Org. Waste Agric.* **2021**, *10*, 29–42. [CrossRef]
16. Ding, Z.; Kheir, A.M.S.; Osama, A.M.A.; Hafez, E.M.; El Shamey, E.A.; Zhou, Z.; Wang, B.; Lin, X.; Ge, Y.; Fahmy, A.E.; et al. A vermicompost and deep tillage system to improve saline-sodic soil quality and wheat productivity. *J. Environ. Manag.* **2021**, *277*, 111388. [CrossRef]
17. Jakubus, M. A comparative study of composts prepared from various organic waste based on biological and chemical parameters. *Agronomy* **2020**, *10*, 869. [CrossRef]
18. Sharma, B.; Sarkar, A.; Singh, P.; Singh, R.P. Agricultural utilisation of biosolids: A review on potential effects on soil and plant grown. *Waste Manag.* **2017**, *64*, 117–132. [CrossRef]
19. Lv, B.; Xing, M.; Yang, J.; Qi, W.; Lu, Y. Chemical and spectroscopic characterisation of water extractable organic matter during vermicomposting of cattle dung. *Bioresour. Technol.* **2013**, *132*, 320–326. [CrossRef]
20. Kleber, M.; Lehmann, J. humic substances extracted by alkali are invalid proxies for the dynamics and functions of organic matter in terrestrial and aquatic ecosystems. *J. Environ. Qual.* **2019**, *48*, 207–216. [CrossRef]
21. Directive 2008/98/EC of the European Parliament and of the Council of 19 November 2008 on Waste and Repealing Certain Directives OJ L 312, 22 November 2008; p. 3. Available online: https://www.legislation.gov.uk/eudr/2008/98 (accessed on 15 December 2021).
22. Ostrowska, A.; Gawliński, S.; Szczubialka, Z. *Methods for Analysis and Evaluation of Soil and Plant Properties*, 1st ed.; IOŚ: Warszawa, Poland, 1991; pp. 158–167. (In Polish)
23. Dziadowiec, H.; Gonet, S. *A Methodological Guide to Soil Organic Matter Research*, 1st ed.; PTG: Warszawa, Poland, 1999; Volume 120, pp. 43–51. (In Polish)
24. Loginov, W.; Wisniewski, W.; Gonet, S.S.; Cieścinska, B. Fractionation of organic carbon based on susceptibility to oxidation. *Pol. J. Soil Sci.* **1987**, *20*, 47–52.
25. Ghani, A.; Dexter, M.; Perrott, K.W. Hot-water extractable carbon in soils: A sensitive measurement for determining impacts of fertilisation, grazing and cultivation. *Soil Biol. Biochem.* **2003**, *35*, 1231. [CrossRef]
26. Orlov, D.S.; Grišina, L.A. *Guide of Humus Chemistry*, 1st ed.; IMU: Moskva, Russia, 1981; pp. 1–272. (In Russian)
27. Mushtaq, M.; Iqbal, M.K.; Khalid, A.; Khan, R.A. Humification of poultry waste and rice husk using additives and its application. *Int. J. Recycl. Org. Waste Agric.* **2019**, *8*, 15–22. [CrossRef]
28. Ozdemir, S.; Dede, G.; Dede, O.H.; Turo, S.M. Composting of sewage sludge with mole cricket: Stability, maturity and sanitation aspects. *Int. J. Environ. Sci. Technol.* **2019**, *16*, 5827–5834. [CrossRef]
29. Wong, J.W.C.; Mak, K.F.; Chan, N.W.; Lam, A.; Fang, M.; Zhou, L.X.; Wu, Q.T.; Lia, X.D. Co-composting of soybean residues and leaves in Hong Kong. *Bioresour. Technol.* **2001**, *76*, 99–106. [CrossRef]
30. Sciubba, L.; Cavani, L.; Grigatti, M.; Ciavatta, C.; Marzadori, C. Relationships between stability, maturity, water-extractable organic matter of municipal sewage sludge composts and soil functionality. *Environ. Sci. Pollut. Res.* **2015**, *22*, 13393–13403. [CrossRef]
31. Ibrahim, M.I.M.; Awad, E.A.M.; Dahdouh, S.M.M.; El-Etr, W.M.T.; Ibrahim, A.S.M. Characterisations of some organic materials sources and analysis of the humic acids extracted from them. *Zagazig J. Agric. Res.* **2019**, *46*, 685–698.
32. Regulation of the Minister of Agriculture and Rural Development of 18 June 2008. *J. Laws* **2008**, *119*, 765. Available online: http://isap.sejm.gov.pl/isap.nsf/DocDetails.xsp?id=wdu20081190765 (accessed on 15 December 2021). (In Polish).
33. Ramnarain, Y.I.; Ansari, A.A.; Ori, L. Vermicomposting of different organic materials using the epigenic earthworm. *Int. J. Recycl. Org. Waste Agric.* **2019**, *8*, 23–36. [CrossRef]
34. Yu, H.; Xie, B.; Khan, R.; Shen, G. The changes in carbon, nitrogen components and humic substances during organic-inorganic aerobic co-composting. *Bioresour. Technol.* **2019**, *271*, 228–235. [CrossRef]
35. Bu, X.; Wang, L.; Ma, W.; Yu, X.; McDowell, W.H.; Ruan, H. Spectroscopic characterization of hot-water extractable organic matter from soils under different vegetation types along an elevation gradient in the Wuyi Mountains. *Geoderma* **2010**, *159*, 139–146. [CrossRef]
36. Li, S.; Li, D.; Li, J.; Li, G.; Zhang, B. Evaluation of humic substances during co-composting of sewage sludge and corn stalk under different aeration rates. *Bioresour. Technol.* **2017**, *245*, 1299–1302. [CrossRef] [PubMed]

37. Azim, K.; Soudi, B.; Boukhari, S.; Perissol, C.; Roussos, S.; Alami, T. Composting parameters and compost quality. *Org. Agric.* **2018**, *8*, 141–158. [CrossRef]
38. Chen, L.; de Haro, M.M.; Moore, A.; Falen, C. The Composting Process. Dairy Compost Production and Use in Idaho; University of Idaho. CIS 1179, 2011. Available online: https://www.extension.uidaho.edu/publishing/pdf/cis/cis1179.pdf (accessed on 15 December 2021).
39. Gomez-Brandon, M.; Lazcano, C.; Dominguez, J. The evaluation of stability and maturity during the composting of cattle manure. *Chemosphere* **2008**, *70*, 436–444. [CrossRef] [PubMed]
40. Singh, Y.K.; Kalamdhad, A.S.; Ali, M.; Kazmi, A.A. Maturation of primary stabilized compost from rotary drum composter. *Resour. Conserv. Recycl.* **2009**, *53*, 386–392. [CrossRef]
41. Antil, R.S.; Raj, D.; Abdalla, N.; Inbushi, K. Physical, chemical and biological parameters for compost maturity assessment: A review. In *Composting for Sustainable Agriculture*; Maheshwari, D., Ed.; Springer: Cham, Switzerland; Berlin/Heidelberg, Germany; New York, NY, USA; Dordrecht, The Netherlands; London, UK, 2014; pp. 83–101. [CrossRef]
42. Asquer, C.; Cappai, G.; Gioannis, G.; Muntoni, A.; Piredda, M.; Spiga, D. Biomass ash reutilization as an additive in the composting process of organic fraction of municipal solid waste. *Waste Manag.* **2017**, *69*, 127–135. [CrossRef]
43. Balachandar, R.; Biruntha, M.; Yuvaraj, A.; Thangaraj, R.; Subbaiya, R.; Govarthanan, M.; Kumar, P.; Karmegam, N. Earthworm intervened nutrient recovery and greener production of vermicompost from Ipomoea staphylina—An invasive weed with emerging environmental challenges. *Chemosphere* **2020**, *263*, 128080. [CrossRef]
44. Raj, D.; Antil, R.S. Evaluation of maturity and stability parameters of composts prepared from agro-industrial wastes. *Bioresour. Technol.* **2011**, *102*, 2868–2873. [CrossRef]
45. Bustamante, M.A.; Alburquerque, J.A.; Restrepo, A.P.; de la Fuente, C.; Paredes, C.; Moral, R.; Bernal, M.P. Co-composting of the solid fraction of anaerobic digestates, to obtain added-value materials for use in agriculture. *Biomass Bioenergy* **2012**, *43*, 26–35. [CrossRef]
46. Alvarenga, P.; Mourinha, C.; Farto, M.; Santos, T.; Palma, P.; Sengo, J.M.C.; Morais, M.C.; Cunha-Queda, C. Quality assessment of a battery of organic wastes and composts using maturity, stability and enzymatic parameters. *Waste Biomass Valoris* **2016**, *7*, 455–465. [CrossRef]
47. Tripathi, D.K.; Singh, S.; Singh, S.; Mishra, S.; Chauhan, D.K.; Dubey, N.K. Micronutrients and their diverse role in agricultural crops: Advances and future prospective. *Acta. Physiol. Plant.* **2015**, *37*, 139. [CrossRef]
48. Rodrigues, L.C.; Puig-Ventosa, I.; López, M.; Martínez, F.X.; Ruiz, A.G.; Bertrán, T.G. The impact of improper materials in biowaste on the quality of compost. *J. Clean Prod.* **2020**, *251*, 119601. [CrossRef]
49. Commission Regulation (EC) No 889/2008. Available online: https://eur-lex.europa.eu/legal-content/EN/TXT/PDF/?uri=CELEX:32008R0889&from=EN (accessed on 15 December 2021).

agriculture

MDPI

Article

Effect of Sulfur-Enriched Vermicompost on the Growth of *Brassica chinensis* L. and the *Spodoptera litura* Fabricius Larvae Feeding

Chang-Jun Fong [1], Yi-Yuan Chuang [2] and Hung-Yu Lai [3,4,*]

1 Master's Program for Plant Medicine and Good Agricultural Practice, National Chung Hsing University, Taichung City 40227, Taiwan; fcj@dragon.nchu.edu.tw
2 Department of Entomology, National Chung Hsing University, Taichung City 40227, Taiwan; chuangyiyu@dragon.nchu.edu.tw
3 Department of Soil and Environmental Sciences, National Chung Hsing University, Taichung City 40227, Taiwan
4 Innovation and Development Center of Sustainable Agriculture, National Chung Hsing University, Taichung City 40227, Taiwan
* Correspondence: soil.lai@nchu.edu.tw; Tel.: +886-4-22840373 (ext. 4406); Fax: +886-4-22851969

Abstract: (1) Background: Vermicompost is enriched with plant essential nutrients and has been shown to suppress the incidence of pests; however, its potential is affected by its food sources. (2) Methods: Earthworms were fed cabbage or pig manure to produce two vermicomposts enriched in sulfur and nutrients, respectively. A pot experiment and a feeding experiment were then conducted to determine whether the application of the vermicomposts and sulfur could increase soil fertility, promote the growth of *Brassica chinensis* L., and inhibit the growth of *Spodoptera litura* Fabricius larvae. (3) Results: The characteristics of the vermicomposts were mainly affected by the food sources, and vermicomposted cabbage was found to have a higher sulfur content than vermicomposted pig manure. The application of the vermicomposts enhanced the concentrations of organic matter and available phosphorus, as well as the exchange concentrations of potassium, cadmium, and magnesium in the soil. Moreover, the growth of and the accumulated phosphorus and sulfur in the *B. chinensis* L. samples significantly increased when the plants were grown in soils treated with the two vermicomposts. Hence, the addition of vermicomposted cabbage and sulfur fertilizers can decrease the relative growth rate, total consumption, efficiency of conversion of ingested food, and relative consumption rate of *S. litura* larvae, possibly due to the increase in leaf sulfur concentration.

Keywords: soil fertility; *Spodoptera litura* Fabricius larvae; sulfur; vermicompost

1. Introduction

Large amounts of agricultural waste are produced by the agricultural activities that are required to meet the needs of the increasing human population. Agricultural waste can be converted into vermicompost (VC) when earthworms and microorganisms cooperate, and, due to the lower temperatures at which vermicomposting takes place, there is generally a greater amount and diversity of microorganisms present during vermicomposting than during composting [1]. Given that higher macronutrient concentrations have been reported in VC compared with compost [2,3], it is not surprising that the application of VC has been demonstrated to preserve and restore soil quality and plant growth [4–6]. In addition to a beneficial effect on plant yield, a number of studies have also reported that VC application induces biological resistance in plants against diseases and pests due to the presence of actinomycetes and antibiotics [7–9]. A meta-analysis conducted by Blouin et al. [10] found that the application of VC enhances commercial crop production, total biomass, shoot biomass, and root biomass by 26%, 13%, 78%, and 57%, respectively. Furthermore, VC has potential as an environmentally friendly alternative for the control of pests and diseases.

Many chemicals used in conventional agriculture to suppress pests and diseases have been shown to decrease the population of beneficial soil organisms and to have negative effects on environmental quality [11]. Therefore, the development of environmentally friendly alternatives, such as VC, is crucial.

Vermicomposting is a degradation process mediated by interactions between earthworms and microorganisms that results in the conversion of organic matter (OM) into VC. Only 5–10% of the OM is metabolized by the earthworms during this process; consequently, VCs have a high OM content [12]. Many studies have demonstrated that the application of VC can enhance aggregate stability and aeration [13–15], improve soil quality, and promote plant growth [4–6]. Since vermicomposting is conducted at ambient temperature, the VC microbial population is richer than that of the raw material [16] and can include nitrogen-fixing bacteria, phosphorus-solubilizing bacteria, mycorrhizal fungi, and actinomycetes [4,17]. This enriched microbial population has led to VC being used as a biological control material to suppress pests, parasitic nematodes, and many diseases [18,19]. The possible mechanisms employed to suppress pest attacks include the release of phenolic and toxic substances, an increase in the number and diversity of active microbes and pathogenic nematodes, and an increase in the availability of nutrients [11,20]. In addition, it has been reported that Brassicaceae family members utilize sulfur (S) to synthesize glucosinolate and suppress the growth of many insects [21,22]. Field and greenhouse experiments have also demonstrated that the application of VC significantly reduced pest damage to tomato and cucumber plants [23].

According to the experimental results of our previous study [24], the characteristics of VC are affected by food supplements, and amendments to VC can improve soil fertility and the growth of pak choi (*Brassica chinensis* L.). Moreover, increasing the soil and leaf S content can decrease the relative growth rate (RGR) of tobacco cutworm (*Spodoptera litura*) larvae. Therefore, in this study, and in accordance with the experimental results of Fong et al. [24], two VCs were selected and then applied to the soil used to grow pak choi. The leaves of mature pak choi were infested with *S. litura* larvae to assess the suppressive potential of the two VCs using four nutritional indexes. The objectives of this study included assessing (I) the effect of the VCs on soil fertility and pak choi growth, and (II) the influence of the different VC and S treatments on the secondary metabolite content, antioxidant capacity, and resistance to *S. litura* larvae of pak choi.

2. Materials and Methods

2.1. VC, Crop, Soil, and Larvae

A combination of two species of epigeic earthworms, red wiggler (*Eisenia andrei* or *Eisenia foetida*) and Indian blue worm (*Perionyx excavates*), were used in this study. Used shiitake mushroom sawdust was used as the primary medium for the growth of the earthworms, and they were fed either pig manure or cabbage. In total, 5.0 kg of used shiitake mushroom sawdust was placed in an opaque rectangular polypropylene box (L 47 cm × W 33 cm × H 18 cm), the moisture content was adjusted to 70–75% by adding deionized water (DI water), and 0.5 kg of earthworms was added. The top of each box was covered with a 32-mesh nylon net to avoid both the escape of earthworms and predation by animals. After one week of incubation, 50 g of fresh pig manure or cabbage was added every two days, and residual food was removed if necessary. The moisture content during vermicomposting was adjusted to 70–75% by weighing the box and adding DI water every two to three days. The supply of food and DI water was stopped on day 53, and the VCs produced were collected seven days later. The VC produced by feeding the earthworms pig manure was termed VPM, and the VC produced by feeding the earthworms cabbage was termed VCM. The two VCs were air-dried and then used in the S treatment and pot experiments described in Sections 2.2 and 2.3, respectively.

Pak choi (*Brassica chinensis* L. var. Chinensis), a leafy vegetable commonly found in Taiwanese markets and whose leaf is usually consumed by *S. litura* larvae, was used as the target crop. The surface layer (0–30 cm) of an important soil series, Erhlin, located in central Taiwan, was selected as the study soil. Soil samples were air-dried, ground, sieved with a 5-mesh stainless sieve, and then used in the pot experiment. Second-instar *S. litura* larvae were bought from the Taiwan Agricultural Chemicals and Toxic Substances Institute, and third-instar *S. litura* larvae were used in the infesting experiments. Two separate experiments, the details of which are given in Sections 2.2 and 2.3, were conducted, and the recommended doses (RDs) of nitrogen (N), phosphoric oxide (P_2O_5), and potassium oxide (K_2O) for pak choi recommended by the Agriculture and Food Agency of the Council of Agriculture were 250, 150, and 180 kg ha^{-1}, respectively.

2.2. Sulfur Treatment Experiment

VPM and VCM were selected because in previous experiments they were found to be the best promoter of pak choi growth and to have the highest content of S, respectively [24]. Higher amounts of S fertilizer and VC were applied in this experiment compared to previous experiments to increase the S content in the soil and in the pak choi, and their effect on decreasing the nutritional indexes of *S. litura* larvae was assessed. The following nine treatments were tested with four replicates each: CK (control): no amendments; CF (1×): chemical fertilizers $CO(NH_2)_2$, $Ca(H_2PO_4)_2{\cdot}H_2O$, and KCl at the RDs; CF (1.5×): the same chemical fertilizers as CF (1×) at 1.5 times the RDs; CF + S (1×): ammonium sulfate ($(NH_4)_2SO_4$), $Ca(H_2PO_4)_2{\cdot}H_2O$, and KCl at the RDs; CF + S (1.5×): the same chemical fertilizers as CF + S (1×) at 1.5 times the RDs; VPM (1×); VPM (1.5×); VCM (1×); and VCM (1.5×). Unlike chemical fertilizers, VC has to be mineralized to release inorganic N so that it can be taken up by plants. Therefore, in the VPM (1×), VPM (1.5×), VCM (1×), and VCM (1.5×) treatments, VC was added at 2.5 times (1×) or 3.75 times (1.5×) the RD for N according to the N content of the two VCs.

2.3. Pot Experiment

The sieved soil samples prepared as outlined in Section 2.1 were homogeneously mixed with the different chemical fertilizers or VCs prepared as described in Section 2.2, and then 1.0 kg of mixture was added to each pot. The pot experiment was conducted in a growth chamber (14 h of light, temperature of 25.16 ± 1.66 °C, and relative humidity of 60.83 ± 17.17%) and 30 pak choi seeds were sown in each pot. The soil moisture during the pot experiment was controlled at 50–70% of the water-holding capacity by weighing and adding DI water every two to three days. Only five pak choi seedlings with similar shoot heights were left for seven days after germination and the others were removed.

For the infesting experiment (described in Section 2.1), two pak choi plants from each treatment were randomly selected and two separate sub-experiments were conducted. In the first sub-experiment, one of the pak choi replicates was infested with four third-instar *S. litura* larvae for one week, and the whole pak choi was covered with 32-mesh nylon during the experiment. In the second sub-experiment, three third-instar *S. litura* larvae were placed on individual Petri dishes and fed with the third and fourth leaves of the second pak choi replicate. The leaves were first washed with DI water, and the stems were placed in 2 mL centrifuge tubes, which were then filled with DI water and sealed with paraffin film. The second sub-experiment lasted for one week with four replicates, and the pak choi leaf was renewed every two days.

After growing for seven weeks, the shoots of the pak choi grown with the different treatments were harvested and washed with tap water and DI water, and then shoot height and fresh weight were determined. The relative chlorophyll content (i.e., the SPAD reading) of the most extended leaf of each pak choi replicate was determined using a chlorophyll meter (SPAD-502, Konica Minolta, Osaka, Japan). Plant tissues were oven-dried at 70 °C for 72 h or freeze-dried for 48 h in preparation for the property analyses.

2.4. VC, Soil, Plant, and Larvae Analyses

The moisture content, pH (w/v = 1/5) [25], electrical conductivity (EC; w/v = 1/5) [26], total nitrogen content (TN) [27], and OM [28] of the two VCs were analyzed. In addition, each VC was digested with nitric acid and perchloric acid ($v/v = 4/1$) [29], filtered through Whatman No. 42 filter paper, and quantitated. The P concentration in the filtered digestants was determined in accordance with Murphy and Riley [30]. The K and S concentrations in the filtered digestants were determined with a flame photometer (Sherwood 410, Sherwood Scientific Ltd., Cambridge, UK) and an ion chromatography system (930 Compact IC Flex, Metrohm, Herisau, Switzerland), respectively. The concentrations of calcium (Ca) and magnesium (Mg) in the digestants were determined using an atomic absorption spectrometer (Z-2000, Hitachi, Tokyo, Japan), and the cadmium (Cd), chromium (Cr), copper (Cu), nickel (Ni), lead (Pb), and zinc (Zn) concentrations in the digestants were determined using an inductively coupled plasma atomic emission spectrometer (ICP-OES Avio 200, Perkin Elmer, MA, USA).

Soil samples were collected after the pot experiment, air-dried, ground, and passed through 10-mesh or 80-mesh sieves according to the properties analyzed. The pH, EC, and OM of the soil samples were analyzed using the methods described above. Other properties analyzed included: concentrations of available N [30], available P [31], and available S [32]; exchangeable concentrations of K, Ca, and Mg [33]; and wet aggregate stability (WAS) [34]. The oven-dried plant tissue was ground, digested with nitric acid and perchloric acid ($v/v = 4/1$) [29], filtered through Whatman No. 42 filter paper, and quantitated. The concentrations of N, P, K, Ca, Mg, and S in the filtered digestant were then determined using the method outlined above. The freeze-dried plant tissue was used to determine the 1,1-diphenyl-2-picrylhydrazyl (DPPH) free radical scavenging ability and the concentrations of total phenolics and total flavonoids in accordance with Hatano et al. [35].

In the sulfur treatment experiment described in Section 2.2, the nutritional indexes used by Farrar et al. [36] and Nawaz et al. [37] were calculated using Equations (1)–(4) to identify the effect of the treatments on the suppression of *S. litura* larval growth.

$$\text{Relative growth rate (RGR)} = \frac{\text{Fresh weight increase of larvae/Initial fresh weight of larvae}}{\text{Period of experiment}} \quad (1)$$

$$\text{Total consumption (TC)} = \text{Initial fresh weight of food} - \text{final fresh weight of food} \quad (2)$$

$$\text{Relative consumption rate (RCR)} = \frac{\text{Food consumption/Initial fresh weight of larvae}}{\text{Period or experiment}} \quad (3)$$

$$\text{Efficiency of conversion of ingested food (ECI)} = \frac{\text{Fresh weight increase of larvae}}{\text{Food consumption}} \times 100 \quad (4)$$

2.5. Statistical Analysis

Statistical analysis was performed using the Statistical Analysis System (SAS) v9.4 software. A one-way analysis of variance (ANOVA) was performed using a generalized linear model (GLM) across the treatments. Fisher's protected least significant difference (LSD) test was used to identify significant differences between means, and $p < 0.05$ denoted statistical significance.

3. Results and Discussion

3.1. The Properties of the Two VCs

The basic properties of the two VCs used in this study are shown in Table 1. Relative to VPM, VCM had a higher pH value, which possibly resulted from the difference in the food source and OM mineralization rate. The release of organic acids (i.e., fulvic acid and humic acid) during the degradation of OM can decrease the pH value [38]; however, the pH of VC increases when salts are released during the degradation of OM [39]. Huang et al. [5] reported that more leachate was produced by earthworms when the food had a high water content. Therefore, the lower EC of VCM compared to that of VPM possibly resulted from cabbage having a higher water content than pig manure, which may have led to more soluble salts leaching out of VCM during vermicomposting. The OM content of VPM and VCM was 68.3% and 70.2%, respectively, and the C/N ratio for both was between 16 and 24. The C/N ratio was regressed as an important index in the assessment of compost maturation and quality [40]. Immobilization is preferred, and a N depression period might have occurred during the degradation of the OM when material with a high C/N ratio was applied [39].

Table 1. The characteristics [1] of two vermicomposts [2].

	pH	EC	Water Content	OM	TN	P_2O_5	K_2O	CaO	MgO	S
		dS m^{-1}				%				
VPM	6.9	5.2	46.8	68.3	2.4	2.5	1.2	4.3	1.0	0.33
VCM	8.5	2.1	34.0	70.2	1.7	1.4	1.3	4.5	0.9	0.51

[1] EC: electrical conductivity; OM: organic matter content; TN: total nitrogen content. [2] VPM: vermicomposted pig manure; VCM: vermicomposted cabbage.

Compared to VCM, VPM had higher concentrations of total N and P_2O_5. In agreement with the findings of Fahey et al. [41], the S content of VCM (0.51%) was higher than that of VPM (0.33%). Cd, Cu, Ni, and Pb were not detectable in the two VCs. Cr and Zn were detected in the two VCs at concentrations of <1 mg kg^{-1} and 10–12 mg kg^{-1}, respectively.

3.2. Sulfur Treatment

3.2.1. The Effect of Sulfur on Soil Properties

The eight additive treatments resulted in a significantly lower soil pH ($p < 0.05$) than the CK treatment, especially the two CF + S treatments; however, all the pH values were in the alkaline range (Table 2). This was due to sulfate ammonium being used to supply N and S in the CF + S treatment and H^+ being secreted by roots and produced during nitrification. In agreement with the results described above, soils amended with VPM and VCM had a lower pH than the CK soil, which was possibly due to the release of organic acids during VC degradation [38,42]. Relative to the CK soil, soils amended with the other eight treatments had significantly ($p < 0.05$) higher EC, particularly those that received the two CF + S treatments, which reached 19–31 mS m^{-1}. This phenomenon was possibly the result of the chemical fertilizer treatments having higher NH_4^+, SO_4^{2-}, K^+, and Cl^- contents. Moreover, the additional H^+ ions in the CF + S treatments (which resulted in the low pH values) possibly acted as exchangeable cations, replacing the exchangeable sites of the soil and thus increasing the EC. As the two VCs were 68–71% OM (Table 1), the addition of different amounts of the two VCs significantly increased the soil OM content from 2.7% (CK treatment) to 3.7–5.0% ($p < 0.05$). High soil OM content is helpful in increasing crop yield; for example, the humic acid released during VC degradation has been shown to promote crop growth and yield [43,44]. The enhancing effect of VC on WAS [13–15] was not observed in this study because the pot experiment was conducted for only a short period of seven weeks.

Table 2. The soil properties [1] after the pot experiment, in which soil was subjected to different treatments [2].

Treatment	pH	EC	OM	WAS	Avail. N	Avail. P	Ex. K	Avail. S	Ex. Ca	Ex. Mg	Ca/Mg
		mS m^{-1}	%		mg kg^{-1}				g kg^{-1}	mg kg^{-1}	Mole Ratio
					Without *Spodoptera litura*						
CK	8.69 ± 0.05 a	8.04 ± 0.28 k	2.78 ± 0.09 d	57.7 ± 5.6 a	13.5 ± 0.0 bc	6.61 ± 1.05 j	45.1 ± 2.8 ef	3.74 ± 1.02 d	3.27 ± 0.04 def	236 ± 8 fg	8.33 ± 0.28 ab
CF (1×)	8.41 ± 0.02 b	11.3 ± 1.7 ghij	2.81 ± 0.08 d	52.6 ± 8.2 abcd	20.2 ± 8.2 b	15.7 ± 1.2 hi	55.5 ± 4.6 ef	3.73 ± 0.81 d	3.14 ± 0.06 gh	217 ± 9 gh	8.69 ± 0.38 a
CF (1.5×)	8.27 ± 0.05 c	13.5 ± 1.5 efg	2.79 ± 0.06 d	51.0 ± 11.8 ab	33.8 ± 16.8 a	20.4 ± 0.4 fg	57.3 ± 3.8 def	4.25 ± 0.86 d	3.16 ± 0.11 fgh	217 ± 9 gh	8.73 ± 0.26 a
CF + S (1×)	8.10 ± 0.02 de	19.1 ± 1.9 c	2.75 ± 0.10 d	58.2 ± 7.0 abc	18.6 ± 10.0 b	14.1 ± 0.7 i	52.3 ± 4.7 ef	107 ± 20 c	3.20 ± 0.05 efg	224 ± 13 fgh	8.60 ± 0.37 ab
CF + S (1.5×)	7.86 ± 0.04 i	29.3 ± 0.6 a	2.80 ± 0.06 d	56.0 ± 8.5 bcd	13.5 ± 4.8 bc	18.5 ± 2.8 gh	47.5 ± 4.8 ef	198 ± 18 b	3.22 ± 0.09 defg	229 ± 8 fgh	8.47 ± 0.25 ab
VPM (1×)	8.12 ± 0.05 d	12.1 ± 2.3 ghi	4.02 ± 0.14 b	53.2 ± 7.7 d	8.45 ± 2.93 c	30.3 ± 0.8 cd	70.1 ± 10.2 cde	5.77 ± 3.10 d	3.55 ± 0.12 b	389 ± 21 cd	5.49 ± 0.23 e
VPM (1.5×)	8.06 ± 0.03 def	16.0 ± 0.6 de	4.20 ± 0.24 b	44.8 ± 3.2 cd	13.5 ± 4.8 bc	42.5 ± 1.5 b	60.8 ± 4.6 cdef	9.53 ± 0.81 d	3.40 ± 0.06 c	409 ± 3 c	5.00 ± 0.11 efg
VCM (1×)	8.26 ± 0.03 c	17.4 ± 0.5 cd	3.71 ± 0.05 c	41.3 ± 2.8 abcd	8.45 ± 2.93 c	34.4 ± 3.1 c	140 ± 15 b	8.53 ± 0.41 d	3.55 ± 0.09 ab	392 ± 14 c	5.44 ± 0.14 e
VCM (1.5×)	8.27 ± 0.10 c	15.5 ± 3.7 def	4.95 ± 0.35 a	50.9 ± 12.2 bcd	13.5 ± 6.7 bc	49.9 ± 5.2 a	240 ± 55 a	6.97 ± 3.61 d	3.67 ± 0.05 a	468 ± 24 a	4.72 ± 0.25 fg
					Infested with *S. litura*						
CK	8.41 ± 0.02 b	11.0 ± 0.4 hij	2.72 ± 0.05 d	62.6 ± 2.1 a	10.1 ± 5.6 c	7.95 ± 0.43 j	43.1 ± 1.1 f	4.90 ± 0.51 d	3.06 ± 0.01 hi	242 ± 10 f	7.60 ± 0.32 cd
CF (1×)	8.25 ± 0.05 c	9.15 ± 0.89 jk	2.74 ± 0.06 d	53.9 ± 6.6 abcd	10.1 ± 3.4 c	14.6 ± 0.4 hi	46.4 ± 2.9 ef	1.86 ± 0.16 d	2.86 ± 0.02 j	213 ± 10 h	8.06 ± 0.37 bc
CF (1.5×)	8.10 ± 0.07 de	10.8 ± 1.4 hij	2.67 ± 0.08 d	59.8 ± 3.8 ab	18.6 ± 5.6 b	23.4 ± 4.0 ef	50.9 ± 5.7 ef	2.06 ± 0.30 d	2.87 ± 0.03 j	231 ± 12 fgh	7.49 ± 0.32 d
CF + S (1×)	7.95 ± 0.02 h	22.3 ± 1.2 b	2.64 ± 0.02 d	59.4 ± 5.5 abc	10.1 ± 3.4 c	14.7 ± 0.3 hi	41.4 ± 1.4 f	114 ± 12 c	3.04 ± 0.10 hi	238 ± 21 f	7.73 ± 0.83 cd
CF + S (1.5×)	7.65 ± 0.02 j	30.2 ± 2.2 a	2.78 ± 0.09 d	49.4 ± 11.2 bcd	20.2 ± 4.8 b	27.0 ± 2.9 de	44.9 ± 3.2 ef	225 ± 20 a	2.95 ± 0.04 ij	235 ± 15 fg	7.57 ± 0.52 cd
VPM (1×)	8.02 ± 0.01 fg	13.0 ± 2.8 fgh	4.17 ± 0.11 b	47.4 ± 8.1 d	10.1 ± 3.4 c	29.9 ± 0.4 d	81.4 ± 22.2 cd	8.85 ± 6.19 d	3.14 ± 0.07 gh	366 ± 13 e	5.15 ± 0.12 ef
VPM (1.5×)	7.98 ± 0.03 gh	10.5 ± 0.3 hij	4.74 ± 0.21 a	48.2 ± 5.1 cd	13.5 ± 0.0 bc	41.4 ± 5.9 b	66.0 ± 1.9 cdef	2.65 ± 0.30 d	3.29 ± 0.14 cde	408 ± 5 c	4.84 ± 0.24 fg
VCM (1×)	8.09 ± 0.05 def	10.1 ± 0.4 ijk	4.10 ± 0.23 b	52.1 ± 10.3 abcd	18.6 ± 3.0 b	27.3 ± 1.2 de	83.2 ± 5.7 c	4.17 ± 1.74 d	3.16 ± 0.03 fgh	370 ± 6 de	5.13 ± 0.06 ef
VCM (1.5×)	8.05 ± 0.02 efg	11.7 ± 0.7 ghi	4.87 ± 0.07 a	50.0 ± 3.7 bcd	10.1 ± 3.4 c	44.2 ± 1.9 b	154 ± 18 b	4.62 ± 0.13 d	3.33 ± 0.05 cd	445 ± 7 b	4.49 ± 0.03 g
F-value	87.6	49.2	108.6	0.78	4.51	77.9	33.1	207	28.7	163	71.9
$F_{17,54,0.95}$	1.82										

[1] Mean ± standard deviation (n = 4). Means within a column followed by the same letters are not significantly different at $p < 0.05$. EC: electrical conductivity; OM: organic matter content; TN: total nitrogen content; WAS: wet aggregate stability. [2] CK: no amendments; CF (1×): $CO(NH_2)_2$, $Ca(H_2PO_4)_2 \cdot H_2O$, and KCl based on the recommended amount (RD); CF (1.5×): the same chemical fertilizers as CF (1×) at 1.5 times the RDs; CF + S (1×): $(NH_4)_2SO_4$, $Ca(H_2PO_4)_2$ H_2O, and KCl at the RDs; CF + S (1.5×): the same chemical fertilizers as CF + S (1×) at 1.5 times the RDs; VPM (1×): vermicomposted pig manure (VPM) at the RD, based on VPM's TN; VPM (1.5×): VPM at 1.5 times the RD, based on VPM's TN; VCM (1×): vermicomposted cabbage (VCM) at the RD, based on VCM's TN; VCM (1.5×): VCM at 1.5 times the RD, based on VCM's TN.

At the end of the pot experiment, the concentration of available N was higher or significantly higher in the soils amended with the CF (1.5×) and CF + S (1.5×) treatments than in the CK soil ($p < 0.05$) (Table 2). This might have been due to a faster release rate and the higher amount of fertilizer applied. The concentration of available P significantly increased from 6.6–8.0 mg kg^{-1} (CK treatment) to 14–50 mg kg^{-1} with the CF, CF + S, and two VC treatments ($p < 0.05$). As the two VCs were 1.4–2.5% P_2O_5 (Table 1), the concentration of available P was higher in soils treated with VPM and VCM than in soils treated with CF and CF + S. Among the nine treatments used, the concentration of available P was less than 10 mg kg^{-1} only in CK-treated soil; this is the level considered sufficient for plant growth [31]. As SO_4^{2-} is released through the dissolution of applied $(NH_4)_2SO_4$, the two CF + S treatments significantly increased the available S content in the soil from 1.8–5.0 mg kg^{-1} (CK and two CF treatments) to 107–225 mg kg^{-1} ($p < 0.05$). In agreement with the findings of Ramawtar et al. [45], the application of the two VCs increased the general available S content in the soil, which was the result of VC mineralization. The available S content in the soils amended with the two VCs was in the range of 2.6–9.6 mg kg^{-1}. The concentrations of exchangeable K, exchangeable Ca, and exchangeable Mg were significantly higher in the soils treated with the two VCs compared with the CK soil ($p < 0.05$), which resulted from the high content of K_2O, CaO, and MgO in the two VCs (Table 1). In soils treated with CK, CF, and CF + S, the mole ratio of exchangeable Ca to exchangeable Mg was 8.3–8.8, which is higher than the value recommended (6.0) for plant growth [46,47]. In soils treated with VPM (1×) and VCM (1×), the mole ratio decreased to a much more suitable level, 5.1–5.5, due to a significant increase in exchangeable Mg content.

3.2.2. The Effect of Sulfur on Pak Choi Growth

Table 3 shows the effects of the nine treatments on different aspects of pak choi growth. It was found that the SPAD reading, shoot height, and fresh weight of pak choi increased or significantly increased after the CF, CF + S, and VC treatments compared to the CK treatment ($p < 0.05$). The pak choi grown in soil that received CF and CF + S treatments also exhibited generally better growth than those grown in soils with VC treatments. The supply of N in the soil has a drastic influence on the growth of short-term crops, and N could be readily released from the urea and $(NH_4)_2SO_4$ present in the CF and CF + S treatments; however, organic N must be mineralized to inorganic N before uptake by plants. Of the two VCs used, the two VPM treatments resulted in better pak choi growth than the two VCM treatments, which was possibly due to the higher TN and P_2O_5 contents (Table 1) and the higher concentrations of other nutrients not analyzed in this study. As a result of the different foods used to produce VPM and VCM, the population of microorganisms and thus the mineralization rates were quite different [48], and this might help explain the experimental results.

Table 3. The SPAD readings, shoot heights, and fresh weights of the *Brassica chinensis* samples grown in soils that received different treatments [1].

Treatment	SPAD Reading	Shoot Height	Fresh Weight
		cm	g Plant^{-1}
	Without *Spodoptera litura*		
CK	6.93 ± 0.51 h	19.0 ± 1.4 hij	2.83 ± 0.29 f
CF (1×)	14.6 ± 1.7 a	25.6 ± 2.8 bcd	20.9 ± 2.5 def
CF (1.5×)	14.0 ± 0.6 ab	26.5 ± 2.5 bc	28.3 ± 4.7 bcd
CF + S (1×)	11.8 ± 1.1 bcdef	23.1 ± 3.6 defg	18.1 ± 1.8 def
CF + S (1.5×)	14.8 ± 1.2 a	30.9 ± 2.5 a	46.0 ± 1.5 ab
VPM (1×)	13.5 ± 1.7 abc	22.2 ± 2.1 efg	11.1 ± 0.5 def
VPM (1.5×)	12.8 ± 1.1 abcd	23.5 ± 1.7 def	22.6 ± 3.0 def

Table 3. *Cont.*

Treatment	SPAD Reading	Shoot Height	Fresh Weight
		cm	g Plant^{-1}
	Without *Spodoptera litura*		
VCM (1×)	10.4 ± 0.3 efg	18.2 ± 1.8 ij	4.91 ± 0.22 ef
VCM (1.5×)	10.8 ± 1.1 defg	18.8 ± 1.4 hij	5.91 ± 0.40 ef
	Infested with *S. litura*		
CK	5.15 ± 0.82 h	16.7 ± 1.7 j	3.05 ± 0.88 f
CF (1×)	12.7 ± 1.8 abcde	25.2 ± 2.8 cd	23.7 ± 0.3 cdef
CF (1.5×)	12.2 ± 1.9 bcdef	24.7 ± 5.1 cde	21.8 ± 3.0 def
CF + S (1×)	12.8 ± 1.8 abcd	23.3 ± 5.4 def	51.3 ± 16.8 a
CF + S (1.5×)	12.7 ± 1.1 abcde	28.4 ± 2.9 ab	44.8 ± 15.8 abc
VPM (1×)	11.3 ± 0.6 cdefg	23.4 ± 2.9 def	21.6 ± 0.2 def
VPM (1.5×)	13.1 ± 1.9 abc	24.8 ± 2.7 cde	26.3 ± 8.9 bcde
VCM (1×)	10.0 ± 0.4 fg	20.4 ± 2.3 ghi	9.42 ± 0.27 def
VCM (1.5×)	9.40 ± 1.34 g	21.3 ± 1.8 fgh	8.00 ± 1.18 def
F-value	10.4	15.1	3.96
$F_{17,54,0.95}$	1.82		

[1] The codes have the same meanings as those in Table 2. Mean ± standard deviation (n = 4); means within a column followed by the same letters are not significantly different at $p < 0.05$.

The concentrations of N, P, K, S, Mg, and Ca were determined in the leaves of pak choi plants grown in the treated soils. Only the N, P, K, and S concentrations were significantly higher in the leaves of pak choi grown in soil treated with CF, CF + S, and the two VCs compared to those grown in CK-treated soil ($p < 0.05$). The CF and CF + S treatments significantly increased the N and P concentrations to 1.8–4.2% and 0.32–0.56%, respectively, compared with CK treatment ($p < 0.05$) (Table 4). These results were due to the higher concentrations of available N and available P in the soils compared with the other treatments (Table 2). As the two VCs were 1.4–2.5% P_2O_5 and the concentration of available P after VC treatment was significantly higher than after CK, CF, and CF + S treatment, the pak choi P concentration also increased to 0.34–0.55% when the different VC treatments were applied. Regarding the S concentration, the leaves of pak choi grown in the soils treated with CF + S (1×), CF + S (1.5×), VCM (1×), and VCM (1.5×) had significantly higher S concentrations than leaves from plants grown in CK-treated soil ($p < 0.05$). The high concentrations of available S in the soil (Table 2) most likely contributed to the highest concentrations of S being recorded in the pak choi grown in soil treated with the two CF + S treatments (0.28–0.48%). The concentration of S in the pak choi grown in soil treated with the two VCM treatments also significantly increased from approximately 0.01–0.03% (CK treatment) to 0.05–0.08% ($p < 0.05$). Even though the infection of *S litura* induced an increase in S content in the leaves of pak choi under VCM treatments, the differences were not significant.

Table 4. The concentrations of nitrogen, phosphorous, potassium, calcium, magnesium, and sulfur in the leaves of *Brassica chinensis* grown in soils that received different treatments [1].

Treatment	N	P	K	Ca	Mg	S
			%			
			Without *Spodoptera litura*			
CK	1.16 ± 0.12 de	0.199 ± 0.083 h	2.97 ± 0.24 ef	3.46 ± 0.25 bcd	0.417 ± 0.023 cde	0.011 ± 0.012 f
CF (1×)	2.52 ± 0.39 b	0.403 ± 0.064 cdef	2.93 ± 0.43 efg	4.15 ± 0.62 ab	0.411 ± 0.027 def	0.040 ± 0.009 def
CF (1.5×)	2.81 ± 0.27 b	0.447 ± 0.125 abcde	2.95 ± 0.04 ef	3.68 ± 0.47 bcd	0.352 ± 0.065 ef	0.037 ± 0.001 def
CF + S (1×)	2.93 ± 0.08 b	0.321 ± 0.034 fg	3.52 ± 0.11 bcd	3.31 ± 0.70 cd	0.399 ± 0.022 def	0.387 ± 0.046 b
CF + S (1.5×)	2.87 ± 0.06 b	0.389 ± 0.113 def	3.78 ± 0.35 ab	3.82 ± 0.70 bc	0.454 ± 0.077 abcd	0.477 ± 0.094 a
VPM (1×)	1.66 ± 0.05 cd	0.475 ± 0.019 abcd	2.61 ± 0.20 fgh	2.30 ± 0.26 f	0.383 ± 0.028 def	0.048 ± 0.007 def

Table 4. *Cont.*

Treatment	N	P	K	Ca	Mg	S
			%			
			Without *Spodoptera litura*			
VPM (1.5×)	1.61 ± 0.34 cde	0.520 ± 0.110 abc	3.05 ± 0.36 def	3.20 ± 0.25 cde	0.445 ± 0.033 bcd	0.038 ± 0.010 def
VCM (1×)	1.23 ± 0.10 de	0.472 ± 0.076 abcd	3.41 ± 0.02 bcde	2.31 ± 0.24 f	0.398 ± 0.012 def	0.063 ± 0.009 de
VCM (1.5×)	1.14 ± 0.10 de	0.482 ± 0.066 abcd	3.29 ± 0.06 bcde	2.93 ± 0.98 def	0.395 ± 0.054 def	0.055 ± 0.002 de
			Infested with *S. litura*			
CK	1.68 ± 0.29 cd	0.268 ± 0.059 gh	3.20 ± 0.21 cde	3.63 ± 0.31 bcd	0.518 ± 0.033 ab	0.029 ± 0.003 ef
CF (1×)	2.70 ± 0.29 b	0.556 ± 0.075 a	3.35 ± 0.42 bcde	3.47 ± 0.38 bcd	0.433 ± 0.076 bcde	0.038 ± 0.007 def
CF (1.5×)	4.12 ± 0.93 a	0.514 ± 0.066 abc	3.58 ± 0.71 bc	4.63 ± 0.13 a	0.499 ± 0.025 abc	0.039 ± 0.010 def
CF + S (1×)	1.89 ± 0.31 c	0.340 ± 0.080 efg	2.94 ± 0.08 ef	3.34 ± 0.33 cd	0.373 ± 0.044 def	0.281 ± 0.021 c
CF + S (1.5×)	3.82 ± 0.88 a	0.489 ± 0.086 abcd	4.13 ± 0.66 a	3.99 ± 0.99 abc	0.534 ± 0.142 a	0.324 ± 0.014 c
VPM (1×)	1.04 ± 0.10 e	0.346 ± 0.035 efg	2.42 ± 0.09 gh	2.29 ± 0.21 f	0.329 ± 0.026 f	0.030 ± 0.005 def
VPM (1.5×)	1.16 ± 0.21 de	0.431 ± 0.054 bcdef	2.35 ± 0.11 h	2.34 ± 0.23 f	0.372 ± 0.019 def	0.031 ± 0.009 def
VCM (1×)	1.39 ± 0.03 cde	0.483 ± 0.044 abcd	3.59 ± 0.11 bc	2.37 ± 0.19 f	0.392 ± 0.037 def	0.073 ± 0.003 d
VCM (1.5×)	1.24 ± 0.12 de	0.541 ± 0.012 ab	3.77 ± 0.20 ab	2.45 ± 0.09 ef	0.413 ± 0.013 def	0.063 ± 0.002 de
F-value	21.5	5.48	7.00	6.73	3.26	90.6
$F_{17,54,0.95}$	1.82					

[1] The meanings of the codes are the same as in Table 2. Mean ± standard deviation (n = 4); means within a column followed by the same letters are not significantly different at $p < 0.05$.

The CF and CF + S treatments resulted in pak choi that had a higher DPPH free radical scavenging ability and total flavonoid content than the other treatments (Table 5); however, the additive treatments did not significantly influence total phenolic content compared with the CK treatment. Total flavonoid content increased from 7.7–12.4 mg-QE g-DW^{-1} in pak choi grown in CK-treated soil to 12.7–28.5 and 16.7–35.3 mg-QE g-DW^{-1} in pak choi grown in CF- and CF + S-treated soil, respectively. This was in contrast to the C–N balance theory [49] and the growth–differentiation balance hypothesis [50], which state that the application of N could inhibit the synthesis of C-containing secondary metabolites. In agreement with the results of previous studies [51,52], all treatments increased the DPPH free radical scavenging ability of pak choi to 58–88% compared with the 41–53% of pak choi grown in CK-treated soil.

Table 5. The concentrations of total phenolics, total flavonoids, and DPPH scavenging ability in the leaves of *Brassica chinensis* grown in soils that received different treatments [1].

Treatment	Total Phenolics	Total Flavonoids	DPPH Scavenging Ability
	mg-GAE g-DW^{-1}	mg-QE g-DW^{-1}	%
	Without *Spodoptera litura*		
CK	5.58 ± 0.00 cdef	12.4 ± 0.0 cde	52.9 ± 0.0 gh
CF (1×)	5.82 ± 0.51 bcde	28.5 ± 9.9 ab	80.3 ± 7.0 abcd
CF (1.5×)	4.55 ± 0.23 fgh	12.7 ± 4.5 cde	63.0 ± 2.8 efg
CF + S (1×)	5.66 ± 0.35 cde	35.3 ± 3.1 a	69.8 ± 10.61 cdef
CF + S (1.5×)	5.41 ± 0.38 def	20.7 ± 1.1 abcde	85.0 ± 7.4 ab
VPM (1×)	5.79 ± 0.05 bcde	8.36 ± 4.72 e	69.0 ± 1.1 cdef
VPM (1.5×)	5.57 ± 0.01 def	9.81 ± 0.36 e	75.1 ± 2.0 abcde
VCM (1×)	5.84 ± 0.26 bcde	10.2 ± 2.2 de	69.4 ± 4.9 cdef
VCM (1.5×)	5.55 ± 0.23 def	9.09 ± 0.36 e	75.9 ± 1.1 abcde
	Infested with *S. litura*		
CK	3.95 ± 0.00 h	7.70 ± 0.00 e	41.4 ± 0.0 h
CF (1×)	6.18 ± 0.10 bcd	27.3 ± 8.7 abc	78.0 ± 3.4 abcde
CF (1.5×)	7.85 ± 1.09 a	25.2 ± 9.7 abcd	77.6 ± 11.9 abcde
CF + S (1×)	6.75 ± 0.11 b	25.3 ± 10.5 abcd	80.6 ± 6.0 abcd
CF + S (1.5×)	5.29 ± 0.29 defg	16.7 ± 5.1 bcde	71.0 ± 2.2 bcdef
VPM (1×)	5.70 ± 0.01 bcde	6.91 ± 0.36 h	83.0 ± 2.1 abc
VPM (1.5×)	6.67 ± 0.51 bc	11.6 ± 2.2 de	87.2 ± 1.9 a
VCM (1×)	4.88 ± 0.06 efgh	6.54 ± 0.73 e	66.9 ± 3.1 defg

Table 5. *Cont.*

Treatment	Total Phenolics	Total Flavonoids	DPPH Scavenging Ability
	mg-GAE g-DW^{-1}	mg-QE g-DW^{-1}	%
VCM (1.5×)	4.22 ± 0.38 gh	7.63 ± 1.09 e	58.8 ± 0.1 fg
F-value	6.31	3.12	5.35
$F_{17,54,0.95}$	1.82		

[1] The meanings of the codes are the same as in Table 2. Mean ± standard deviation (n = 4); means within a column followed by the same letters are not significantly different at $p < 0.05$.

There were no statistically significant differences in the concentrations of the six essential elements assessed, the total phenolic content, the total flavonoid content, and the DPPH free radical scavenging ability of the pak choi infested with *S. litura* larvae compared with those not infested. Some plants can avoid being consumed by insects by lowering their nutrient concentrations [53]. The activation and strength of this defense mechanism have been determined not only by testing the saliva composition of insects [54,55] but also by detecting enzymes in insect saliva [56]. Here, infesting *S. litura* larvae did not significantly affect the antioxidant ability or secondary metabolite content of pak choi in general. Since pak choi grown in soil treated with VCM has a higher S content in its leaves (Table 4), another secondary metabolite, glucosinolate, might be responsible for the defense mechanism of pak choi [21].

3.2.3. The Effect of Sulfur on *S. litura* Larvae

The *S. litura* larvae that infested the pak choi grown in the soils treated with CF + S (1×), VCM (1×), and VCM (1.5×) treatments had significantly lower RGRs than those that infested the pak choi grown in the soils that received the other treatments, except for CK ($p < 0.05$) (Figure 1). This aligns with the findings presented in Section 3.2.2, which revealed that the CF + S and VCM treatments increased the S content in the leaves of pak choi and therefore inhibited the growth of *S. litura* larvae. The larval RGR decreased when the larvae were fed pak choi grown in soil that received the VCM treatments; the RGR was 10.4–11.3 mg mg^{-1} d^{-1} when the soil was treated with CF and 7.7–8.9 mg mg^{-1} d^{-1} when the soil was treated with VCM. Although the S could inhibit the growth of larvae and the CF + S (1.5×) treatment resulted in the highest S content in the leaves of pak choi among treatments, the resultant RGR was not the lowest. In addition to S, N might also have an influence on the RGR of *S. litura* larvae. The above phenomenon possibly resulted from the higher N content in the leaves resulting from the CF + S (1.5×) treatment, which may have promoted the growth of larvae, although the higher S content could have inhibited growth. Apart from the VCM (1×) and VCM (1.5×) treatments, the CK treatment also resulted in a lower RGR compared with other treatments. *S. litura* larvae require N-rich foods for growth [57,58]. Thus, the lower RGR associated with plants from CK-treated soil may have been due to the N content being insufficient to support *S. litura* larvae growth.

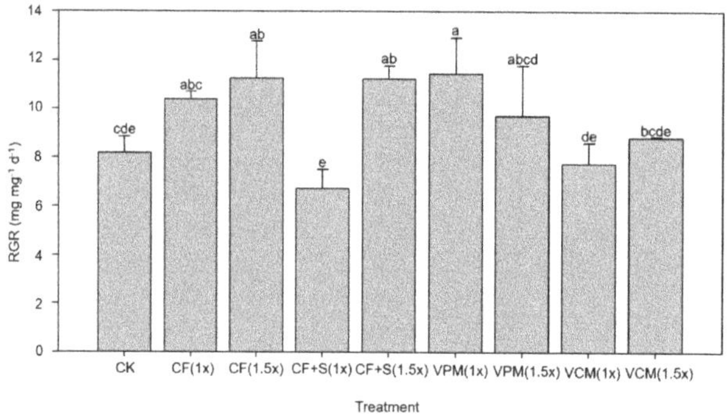

Figure 1. The relative growth rate (RGR) of the *Spodoptera litura* larvae were assessed in the sulfur treatment test. (The meanings of the codes are the same as in Table 2. Means within a column followed by the same letter are not significantly different at $p < 0.05$.)

3.2.4. Short-Term Feeding Experiment

In addition to directly infesting pak choi with *S. litura* larvae, a short-term feeding experiment was conducted for one week, as described in Section 2.3. Three third-instar *S. litura* larvae were grown in individual Petri dishes and then fed two pak choi leaves from plants grown in soils that received different treatments with four replicates. After the one-week experiment, the TC, ECI, RGR, and RCR were calculated using the change in fresh weight of larvae and fresh weight of leaves determined during the experiment, using Equations (1)–(4) [58].

It was found that larvae fed with pak choi grown in soils treated with CK, VCM (1×), and VCM (1.5×) had lower or significantly lower ($p < 0.05$) TC and ECI compared to those fed with pak choi grown in soils that received other treatments (Figure 2). This possibly resulted from the higher leaf S content found in the plants grown in soils that received the two VCM treatments and the lower N content in the plants grown in CK-treated soil, as illustrated in Section 3.2.3. Although the highest S content was observed after the two CF + S treatments, the TC and ECI of the larvae fed the resulting leaves were not the lowest recorded. However, the TC resulting from the two CF + S treatments was lower than that resulting from the two CF treatments, which revealed that raising the S content in the feeding leaves of pak choi decreased larvae consumption. Nawaz et al. [37] recently showed that the ECI based on the dry weight of *S. litura* larvae fed okra was 30–60%, which was higher than that in this study and possibly resulted from the difference in the food source and in the experimental period.

In addition to the TC and ECI, the RCR and RGR of larvae that were fed leaves from pak choi grown in soils treated with CK, VCM (1×), and VCM (1.5×) were generally lower than those resulting from other treatments (Figure 2). In agreement with the TC and ECI results, the CF + S (1×) and CF + S (1.5×) treatments significantly decreased the RCR and RGR compared with the CF (1×) and CF (1.5×) ($p < 0.05$) treatments. A previous study reported that the RCR and RGR of second-instar *S. litura* larvae fed cabbage and okra were 2–3 mg mg^{-1} d^{-1} and 0.6–0.9 mg mg^{-1} d^{-1}, respectively [37]. The higher *S. litura* larval RGR, 3–25 mg mg^{-1} d^{-1}, found in this study possibly resulted from the difference in the food source and the age of the larvae used.

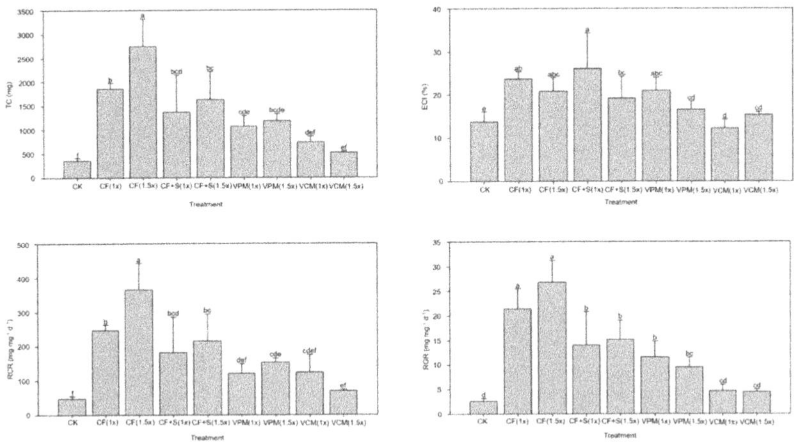

Figure 2. The total consumption (TC), efficiency of conversion of ingested food (ECI), relative consumption rate (RCR), and relative growth rate (RGR) of *Spodoptera litura* larvae in the short-term feeding trial. (The meanings of the codes are the same as in Table 2. Means within a column followed by the same letters are not significantly different at $p < 0.05$.)

3.2.5. The Potential of the VCs to Suppress the Growth of *S. litura* Larvae

The results presented in Sections 3.2.3 and 3.2.4 show that *S. litura* larval growth was determined not only by the S content in the pak choi leaves but also by the N content. This finding is identical to those of previous studies [57,58]. A linear regression was developed using the total N taken up through the consumption of pak choi (x) and the corresponding fresh weight increase in *S. litura* larvae (y) for the non-S treatments (i.e., CK, CF, and VPM treatments) (Figure 3). The theoretical values for fresh weight increase in the S treatments (i.e., CF + S and VCM treatments) were obtained using the total N intake by the *S. litura* larvae and the regressing equation (y = 37.081x + 40.976). The actual values for fresh weight increase in the two CF + S and two VCM treatments were all 16–35% lower than the theoretical values. This revealed that the leaves of the pak choi grown in the S-treated soils (i.e., with CF + S and VCM treatments) had a higher S content, which decreased the food intake and growth of the *S. litura* larvae. High S content in Brassicaceae family members could promote glucosinolate synthesis and thus strengthen the defense against insects [21,22]; the experimental results of this study support this notion. Besides S, enrichment in the number and diversity of microorganisms [18,19] or the release of toxic substances from VC [11,20] may also have contributed to suppressing the growth of *S. litura* larvae; however, these mechanisms were not evidenced in this study using current data. Nevertheless, this study has demonstrated the potential of VCM for the suppression of *S. litura* larval growth.

The CF treatment was included in this study because chemical fertilizers are commonly used in conventional agriculture, and the effects of the CF + S and two VC treatments were compared with those of the CF (1×) treatment. Different relative nutritional indexes were lower after the CF + S, VPM, and VCM treatments than after the CF treatment (Table 6). The two VCM treatments had significantly ($p < 0.05$) lower relative nutritional indexes among the CF, CF + S, and VC treatments, which possibly resulted from the higher S content in the leaves of the pak choi. Due to the low mobility of S in the plant, approximately 90–94% of S accumulates in the old leaves of plants [21]. Two major organic S-containing amino acids, cysteine and methionine, are synthesized using SO_4^{2-} taken up from the soil and are affected by plant maturity [22]; therefore, the parts of the plant fed to the *S. litura* larvae might have also influenced their growth. In the short-term feeding experiment (Section 3.2.4), the third and fourth leaves of the pak choi were fed to the larvae; however, the entire plant was used as the food source in the infesting experiment (Section 3.2.3).

Although the leaves consumed by the larvae might have been quite different, the results revealed that raising the S content in the leaves could restrict the growth of *S. litura* larvae.

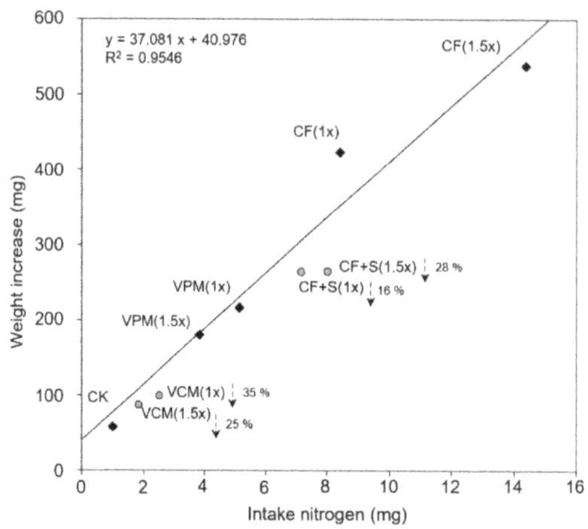

Figure 3. The relationship between intake nitrogen weight (x) and weight increase (y) in *Spodoptera litura* larvae in the short-term feeding trial. (The meanings of the codes are the same as in Table 2.)

Table 6. The relative nutritional indexes [1,2] of *Spodoptera litura* larvae in the short-term feeding experiment.

Treatment	TC	ECI	RCR	RGR
	mg	%	mg mg^{-1} d^{-1}	
CF (1×)	1.000 b	1.000 ab	1.000 ab	1.000 a
CF (1.5×)	1.460 a	0.888 abc	1.410 a	1.220 a
CF + S (1×)	0.812 bc	1.010 a	0.720 bc	0.559 b
CF + S (1.5×)	0.968 b	0.743 bcd	0.979 ab	0.624 b
VPM (1×)	0.697 bcd	0.738 bcd	0.688 bc	0.514 b
VPM (1.5×)	0.609 bcd	0.721 cd	0.603 bc	0.426 bc
VCM (1×)	0.471 cd	0.484 d	0.473 c	0.236 c
VCM (1.5×)	0.337 d	0.594 d	0.347 c	0.212 c
F-value	10.9	4.26	7.23	18.7
$F_{17,54,0.95}$	2.31			

[1] TC: total consumption; ECI: efficiency of conversion of ingested food; RCR: relative consumption rate; RGR: relative growth rate. [2] The meanings of the codes are the same as in Table 2. The relative nutritional indexes = nutritional indexes of each treatment/nutritional indexes of CF (1×). Means within a column followed by the same letters are not significantly different at $p < 0.05$.

4. Conclusions

Agricultural waste can be recycled into vermicompost through the interaction of earthworms and microorganisms. In this study, it has been shown that vermicompost properties can be determined by the food source and that adding vermicompost to soil increases soil fertility and pak choi growth. Moreover, it was found that adding vermicomposted cabbage reduces the growth indexes of *S. litura* larvae consuming pak choi, which is possibly due to the higher S content of vermicomposted cabbage and pak choi leaves. Although the potential of vermicomposted cabbage to reduce the growth of *S. litura* larvae was demonstrated in this study, the presence of parasites and pathogens which may affect human health was not considered in this study. Other mechanisms besides S content must be examined and confirmed in further studies.

Author Contributions: Conceptualization, C.-J.F. and H.-Y.L.; methodology, C.-J.F., Y.-Y.C. and H.-Y.L.; resources and project administration, H.-Y.L.; investigation, C.-J.F.; writing—original draft preparation, C.-J.F. and H.-Y.L.; writing—review and editing, H.-Y.L. All authors have read and agreed to the published version of the manuscript.

Funding: The authors would like to thank the Ministry of Science and Technology of the Republic of China for financially supporting this research under Contract No. MOST 110-2313-B-005-022 and the Ministry of Education of the Republic of China for partial support through the Higher Education Sprout Project.

Acknowledgments: We would like to thank the members of the Soil Survey and Remediation Laboratory in the Department of Soil and Environmental Sciences at National Chung Hsing University for their assistance during the pot experiments and analyses.

Conflicts of Interest: The authors declare no conflict of interest.

References

1. Hernandez, T.; Chocano, C.; Moreno, J.L.; Garcia, C. Towards a more sustainable fertilization: Combined use of compost and inorganic fertilization for tomato cultivation. *Agric. Ecosyst. Environ.* **2014**, *196*, 178–184. [CrossRef]
2. El-Haddad, M.E.; Zayed, M.S.; El-Sayed, G.A.; Abd EL-Satar, A.M. Efficiency of compost and vermicompost in supporting the growth and chemical constituents of *salvia officinalis* L. cultivated in sand soil. *Int. J. Recycl. Org. Waste Agric.* **2020**, *9*, 49–59.
3. Joshi, T.N.; Nepali, D.B.; Sah, R.; Bhattarai, T.; Midmore, D.J. A comparison of composting and vermicomposting for the disposal of poultry waste. *Anim. Prod. Sci.* **2020**, *60*, 986–992. [CrossRef]
4. Sinha, R.K. Earthworms: The miracle of nature (Charles Darwin's 'unheralded soldiers of mankind & farmer's friends'). *Environmentalist* **2009**, *29*, 339.
5. Huang, K.; Xia, H.; Cui, G.; Li, F. Effects of earthworms on nitrification and ammonia oxidizers in vermicomposting systems for recycling of fruit and vegetable wastes. *Sci. Total Environ.* **2017**, *578*, 337–345. [CrossRef] [PubMed]
6. Sharma, K.; Garg, V.K. Comparative analysis of vermicompost quality produced from rice straw and paper waste employing earthworm *Eisenia foetida* (Sav.). *Bioresour. Technol.* **2018**, *250*, 708–715. [CrossRef] [PubMed]
7. Arancon, N.Q.; Edwards, C.A.; Bierman, P.; Metzger, J.D.; Lucht, C. Effects of vermicomposts produced from cattle manure, food waste and paper waste on the growth and yield of peppers in the field. *Pedobiologia* **2005**, *49*, 297–306. [CrossRef]
8. Edwards, C.A.; Arancon, N.; Greytak, S. Effects of vermicompost teas on plant growth and disease. *Biocycle* **2006**, *47*, 28–31.
9. Munroe, G. *Manual of on-Farm Vermicomposting and Vermiculture*; Publication of Organic Agriculture Centre of Canada: Truro, NS, Canada, 2007.
10. Blouin, M.; Barrere, J.; Meyer, N.; Lartigue, S.; Barot, S.; Mathieu, J. Vermicompost significantly affects plant growth. A meta-analysis. *Agron. Sust. Dev.* **2019**, *39*, 34. [CrossRef]
11. Yatoo, A.M.; Ali, M.N.; Baba, Z.A.; Hassan, B. Sustainable management of diseases and pests in crops by vermicompost and vermicompost tea. A review. *Agron. Sustain. Dev.* **2021**, *41*, 7. [CrossRef]
12. Edwards, C.A.; Dominguez, J. Biology and ecology of earthworm species used for vermicomposting. In *Vermiculture Technology: Earthworms, Organic Waste and Environmental Management*; Edwards, C.A., Arancon, N.Q., Sherman, R.L., Eds.; CRC Press Publishing: Boca Raton, FL, USA, 2010; pp. 25–37.
13. Blouin, M.; Hodson, M.E.; Delgado, E.A.; Baker, G.; Brussaard, L.; Butt, K.R.; Dai, J.; Dendooven, L.; Peres, G.; Tondoh, J.E.; et al. A review of earthworm impact on soil function and ecosystem services. *Eur. J. Soil Sci.* **2013**, *64*, 161–182. [CrossRef]
14. Zhu, F.; Hou, J.T.; Xue, S.G.; Wu, C.; Wang, Q.L.; Hartley, W. Vermicompost and gypsum amendments improve aggregate formation in bauxite residue. *Land Degrad. Dev.* **2017**, *28*, 2109–2120. [CrossRef]
15. Liu, M.L.; Wang, C.; Wang, F.Y.; Xie, Y.J. Vermicompost and humic fertilizer improve coastal saline soil by regulating soil aggregates and the bacterial community. *Arch. Agron. Soil Sci.* **2019**, *65*, 281–293. [CrossRef]
16. Domínguez, J.; Aira, M.; Gómez-Brandón, M. Vermicomposting: Earthworms enhance the work of microbes. In *Microbes at Work*; Insam, H., Franke-Whittle, I., Goberna, M., Eds.; Springer: Berlin/Heidelberg, Germany, 2010; pp. 93–114.
17. Yatoo, A.M.; Rasool, S.; Ali, S.; Majid, S.; Rehman, M.U.; Ali, M.N.; Eachkoti, R.; Rasool, S.; Rashid, S.M.; Farooq, S. Vermicomposting: An ecofriendly approach for recycling/management of organic wastes. In *Bioremediation and Biotechnology*; Springer: Cham, Switerland, 2020; pp. 167–187.
18. Edwards, C.A.; Subler, S. Comparison of vermicomposting and composting. *Biocycle* **1997**, *38*, 57–59.
19. Basco, M.J.; Bisen, K.; Keswani, C.; Singh, H.B. Biological management of Fusarium wilt of tomato using biofortified vermicompost. *Mycosphere* **2017**, *8*, 467–483. [CrossRef]
20. Sedaghatbaf, R.; Samih, M.A.; Zohdi, H.; Zarabi, M. Vermicomposts of different origins protect tomato plants against the sweetpotato whitefly. *Econ. Entomol.* **2018**, *111*, 146–153. [CrossRef]
21. Falk, K.L.; Tokuhisa, J.G.; Gershenzon, J. The effect of sulfur nutrition on plant glucosinolate content: Physiology and molecular mechanisms. *Plant Biol.* **2007**, *9*, 573–581. [CrossRef]

22. Ahuja, I.; Rohloff, J.; Bones, A.M. Defence mechanisms of Brassicaceae: Implications for plant-insect interactions and potential for integrated pest management. A review. *Agron. Sustain. Dev.* **2010**, *30*, 311–348. [CrossRef]
23. Yardim, E.N.; Arancon, N.Q.; Edwards, C.A.; Oliver, T.J.; Byrne, R.J. Suppression of tomato hornworm (*Manduca quinquemaculata*) and cucumber beetles (*Acalymma vittatum* and *Diabotrica undecimpunctata*) populations and damage by vermicomposts. *Pedobiologia* **2006**, *50*, 23–29. [CrossRef]
24. Fong, C.J.; Chuang, Y.Y.; Lai, H.Y. Effects of amendment with various vermicomposts on the soil fertility, growth of *Brassica chinensis* L., and resistance of *Spodoptera litura* Fabricius larvae. *Sustainability* **2021**, *13*, 9441. [CrossRef]
25. Thomas, G.W. Soil pH and soil acidity. In *Methods of Soil Analysis: Part 3, SSSA Book Series No. 5, SSSA and ASA*; Sparks, D.L., Page, A.L., Helmke, P.A., Loeppert, R.H., Soltanpour, P.N., Tabatabai, M.A., Johnston, C.T., Sumner, M.E., Eds.; SSSA Inc.: Madison, WI, USA; ASA Inc.: Madison, WI, USA, 1996; pp. 475–490.
26. Rhoades, J.D. Salinity: Electrical conductivity and total dissolved solids. In *Methods of Soil Analysis. Part 3. Chemical Methods*; Sparks, D.L., Page, A.L., Helmke, P.A., Loeppert, R.H., Soltanpour, P.N., Tabatabai, M.A., Johmston, C.T., Sumner, M.E., Eds.; SSSA Inc.: Madison, WI, USA; ASA Inc.: Madison, WI, USA, 1996; pp. 417–435.
27. Bremner, J.M. Nitrogen-Total. In *Methods of Soil Analysis. Part 3. Chemical Methods*; Sparks, D.L., Page, A.L., Helmke, P.A., Loeppert, R.H., Soltanpour, P.N., Tabatabai, M.A., Johmston, C.T., Sumner, M.E., Eds.; SSSA Inc.: Madison, WI, USA; ASA Inc.: Madison, WI, USA, 1996; pp. 1085–1121.
28. Nelson, D.W.; Sommers, L.E. Total carbon, organic carbon, and organic matter. In *Methods of Soil Analysis. Part 3. Chemical Methods*; Sparks, D.L., Page, A.L., Helmke, P.A., Loeppert, R.H., Soltanpour, P.N., Tabatabai, M.A., Johmston, C.T., Sumner, M.E., Eds.; SSSA Inc.: Madison, WI, USA; ASA Inc.: Madison, WI, USA, 1996; pp. 961–1010.
29. Jones, J.B.; Case, V.W. Sampling, handling and analyzing plant tissue samples. In *Soil Testing and Plant Analysis*, 3rd ed.; Westerman, R.L., Ed.; SSSA Inc.: Madison, WI, USA, 1990; pp. 389–427.
30. Murphy, J.; Riley, J.P. A modified single solution method for the determination of phosphate in natural waters. *Anal. Chim. Acta* **1962**, *27*, 31–36. [CrossRef]
31. Sparks, D.L.; Page, A.L.; Helmke, P.A.; Loeppert, R.H.; Soltanpour, P.N.; Tabatabai, M.A.; Johmston, C.T.; Sumner, M.E. (Eds.) *Methods of Soil Analysis. Part 3. Chemical Methods*; SSSA Inc.: Madison, WI, USA; ASA Inc.: Madison, WI, USA, 1996; pp. 869–919.
32. Tabatabai, M.A. Sulfur. In *Methods of Soil Analysis. Part 3. Chemical Methods*; Sparks, D.L., Page, A.L., Helmke, P.A., Loeppert, R.H., Soltanpour, P.N., Tabatabai, M.A., Johmston, C.T., Sumner, M.E., Eds.; SSSA Inc.: Madison, WI, USA; ASA Inc.: Madison, WI, USA, 1996; pp. 921–960.
33. Sumner, M.E.; Miller, W.P. Cation exchange capacity and exchange coefficients. In *Methods of Soil Analysis. Part 3. Chemical Methods*; Sparks, D.L., Page, A.L., Helmke, P.A., Loeppert, R.H., Soltanpour, P.N., Tabatabai, M.A., Johmston, C.T., Sumner, M.E., Eds.; SSSA Inc.: Madison, WI, USA; ASA Inc.: Madison, WI, USA, 1996; pp. 1201–1230.
34. Kemper, W.D.; Rosenau, R.C. Aggregate stability and size distribution. In *Methods of Soil Analysis: Part 1. Physical and Mineralogica Methods*; Klute, A., Campbell, G.S., Jackson, R.D., Mortland, M.M., Nielsen, D.R., Eds.; SSSA Inc.: Madison, WI, USA; ASA Inc.: Madison, WI, USA, 1986; pp. 425–442.
35. Hatano, T.; Kagawa, H.; Yasuhara, T.; Okuda, T. Two new flavonoids and other constituents in licorice root: Their relative astringency and radical scavenging effects. *Chem. Pharm. Bull.* **1988**, *36*, 2090–2097. [CrossRef]
36. Farrar, R.R.; Barbour, J.D.; Kennedy, G.G. Quantifying food consumption and growth in insects. *Ann. Entomol. Soc. Am.* **1989**, *82*, 593–598. [CrossRef]
37. Nawaz, A.; Ali, H.; Sufyan, M.; Gogi, M.D.; Arif, M.J.; Ali, A.; Qasim, M.; Islam, W.; Ali, N.; Bodla, I.; et al. In-vitro assessment of food consumption, utilization indices and losses promises of leafworm, *Spodoptera litura* (Fab.), on okra crop. *Asia Pac. Entomol.* **2020**, *23*, 60–66. [CrossRef]
38. Edwards, C.A. Vermicomposting organic wastes: A review. *Soil Zool. Sustain. Dev. 21st Century* **2004**, *1*, 369–395.
39. Negi, R.; Suthar, S. Degradation of paper mill wastewater sludge and cow dung by brown-rot fungi *Oligoporus placenta* and earthworm (*Eisenia foetida*) during vermicomposting. *Clean. Prod.* **2018**, *201*, 842–852. [CrossRef]
40. Moral, R.; Paredes, C.; Bustamante, M.A.; Marhuenda-Egea, F.; Bernal, M.P. Utilisation of manure composts by high-value crops: Safety and environmental challenges. *Bioresour. Technol.* **2009**, *100*, 5454–5460. [CrossRef] [PubMed]
41. Fahey, J.W.; Zalcmann, A.T.; Talalay, P. The chemical diversity and distribution of glucosinolates and isothiocyanates among plants. *Phytochemistry* **2001**, *56*, 5–51. [CrossRef]
42. Kaviraj, S.S. Municipal solid waste management through vermicomposting employing exotic and local species of earthworms. *Bioresour. Technol.* **2003**, *90*, 169–173. [CrossRef]
43. Maji, D.; Misra, P.; Singh, S.; Kalra, A. Humic acid rich vermicompost promotes plant growth by improving microbial community structure of soil as well as root nodulation and mycorrhizal colonization in the roots of *Pisum sativum*. *Appl. Soil Ecol.* **2017**, *110*, 97–108. [CrossRef]
44. Scaglia, B.; Nunes, R.R.; Rezende, M.O.O.; Tambone, F.; Adani, F. Investigating organic molecules responsible of auxin-like activity of humic acid fraction extracted from vermicompost. *Sci. Total Environ.* **2016**, *562*, 289–295. [CrossRef] [PubMed]
45. Shivran, A.C.; Yadav, B.L. Effect of NP fertilizers, vermicompost and sulphur on growth, yield and quality of clusterbean *Cymopsis tetragonoloba* (L.) and their residual effect on grain yield of succeeding wheat *Triticum aestivum* (L). *Legum. Res.* **2013**, *36*, 74–78.
46. Kopittke, P.M.; Menzies, N.J. A review of the use of the basic cation saturation ratio the ideal soil. *Soil Sci. Soc. Am. J.* **2007**, *71*, 259–265. [CrossRef]

47. Osemwota, I.O.; Omued, J.A.I.; Ogboghodo, A.I. Effect of calcium/magnesium ratio in soil on magnesium availability, yield, and yield components of maize. Comm. *Soil Sci. Plant Anal.* **2007**, *38*, 2849–2860. [CrossRef]
48. Hussain, N.; Singh, A.; Saha, S.; Venkata, S.K.M.; Bhattacharyya, P.; Bhattacharya, S.S. Excellent N-fixing and P-solubilizing traits in earthworm gut-isolated bacteria: A vermicompost based assessment with vegetable market waste and rice straw feed mixtures. *Bioresour. Technol.* **2016**, *222*, 165–174. [CrossRef] [PubMed]
49. Bryant, J.P.; Chapin, F.S.; Klein, D.R. Carbon nutrient balance of boreal plants in relation to vertebrate herbivory. *Oikos* **1983**, *40*, 357–368. [CrossRef]
50. Herms, D.A.; Mattson, W.J. The dilemma of plants: To grow or defend. *Q. Rev. Biol.* **1992**, *67*, 283–335. [CrossRef]
51. Lakhdar, A.; Falleh, H.; Ouni, Y.; Oueslati, S.; Debez, A.; Ksouri, R.; Abdelly, C. Municipal solid waste compost application improves productivity, polyphenol content, and antioxidant capacity of *Mesembryanthemum edule*. *J. Hazard. Mater.* **2011**, *191*, 373–379. [CrossRef] [PubMed]
52. Wang, D.H.; Shi, Q.H.; Wang, X.F.; Wei, M.; Hu, J.Y.; Liu, J.; Yang, F.J. Influence of cow manure vermicompost on the growth, metabolite contents, and antioxidant activities of Chinese cabbage (*Brassica campestris* ssp chinensis). *Biol. Fertil. Soils* **2010**, *46*, 689–696. [CrossRef]
53. Hansen, A.K.; Moran, N.A. The impact of microbial symbionts on host plant utilization by herbivorous insects. *Mol. Ecol.* **2014**, *23*, 1473–1496. [CrossRef]
54. Sambangi, P.; Rani, P. Induction of phenolic acids and metals in *Arachis hypogea* L. plants due to feeding of three Lepidopteran pests. *Arthropod-Plant Interact.* **2013**, *7*, 517–525. [CrossRef]
55. Alba, J.M.; Glas, J.J.; Schimmel, B.C.J.; Kant, M.R. Avoidance and suppression of plant defenses by herbivores and pathogens. *Plant Interact.* **2011**, *6*, 221–227. [CrossRef]
56. Kant, M.R.; Jonckheere, W.; Knegt, B.; Lemos, F.; Liu, J.; Schimmel, B.C.J.; Villarroel, C.A.; Ataide, L.M.S.; Dermauw, W.; Glas, J.J.; et al. Mechanisms and ecological consequences of plant defence induction and suppression in herbivore communities. *Ann. Bot.* **2015**, *115*, 1015–1051. [CrossRef]
57. Deans, C.A.; Sword, G.A.; Behmer, S.T. Revisiting macronutrient regulation in the polyphagous herbivore *Helicoverpa zea* (Lepidoptera: Noctuidae): New insights via nutritional geometry. *Insect Physiol.* **2015**, *81*, 21–27. [CrossRef]
58. Schoonhoven, L.; Van Loon, J.; Dicke, M. *Insect-Plant Biology*; Oxford University Press Publishing: Oxford, UK, 2005; pp. 48–127.

agriculture

MDPI

Article

Rapid, Clean, and Sustainable Bioprocessing of Toxic Weeds into Benign Organic Fertilizer

Dhandapani Banupriya [1], Tabassum-Abbasi [2,*], Tasneem Abbasi [1] and Shahid Abbas Abbasi [1]

1 Centre for Pollution Control & Environmental Engineering, Pondicherry University, Puducherry 605 014, India
2 School of Engineering, Sustainability Cluster, University of Petroleum and Energy Studies, Dehradun 248 007, India
* Correspondence: tab.abbasi@gmail.com

Abstract: A recent report in this journal from these authors, which shows that vermicomposting transforms a toxic weed such as lantana into a benign organic fertilizer, can be of practical utility only if processes can be developed for rapid, inexpensive, and sustainable vermicomposting of these weeds. This paper describes attempts leading to such a process for the vermicomposting of toxic and allelopathic weeds lantana (*Lantana camara*), parthenium (*Parthenium hysterophorus*), and ipomoea (*Ipomoea carnea*). For it, the 'high-rate vermicomposting' concept was employed due to which the weeds could be used for vermicomposting directly in each case without the need for pre-composting or any other form of pretreatment. The manure worm *Eisenia fetida*, which had been cultured on cowdung as feed and habitat, was slow to adapt to the weed-feed but survived and then began to thrive, in all the three weeds, enabling the weeds' sustained and efficient vermicomposting throughout the 16 month's uninterrupted operation of the vermireactors. In all cases the extent of vermicast production per unit time showed a rising trend, indicating that the rate of vermicomposting was set to rise further with time. The vermicomposting was found to accompany a 50 ± 10% loss of organic carbon of each weed with a 50 ± 10% increase in the concentration of total nitrogen as also the weed's additional mineralization. The combined effect was a significant lowering of the carbon-nitrogen ratio, and enrichment of all major, medium, and trace nutrients in the vermicomposts relative to their parent substrates. The findings establish that sustained, direct, and rapid transformation to organic fertilizers of even toxic and allelopathic weeds can be accomplished with the high-rate vermicomposting paradigm.

Keywords: toxic weeds; high-rate vermicomposting; organic fertilizer; *Lantana*; *Ipomoea*; *Parthenium*

1. Introduction

Parthenium *(Parthenium hysterophorus)*, ipomoea *(Ipomoea carnea)*, and lantana *(Lantana camara)* are among three of the world's most pernicious and intransigent of weeds [1–3]. These weeds can be seen growing profusely in open lands, in and around agricultural farms, roadsides, wetlands, and parks [2,4]. They have been invading even forests; for instance, lantana has covered about 87,000 km^2 of forests in India alone and its global invasion potential has been estimated as 11 million km^2 [5]. The estimates of parthenium colonization are even more grim; as much as 350,000 km^2 of land in India has been overtaken by parthenium [6]. Worse, all three weeds are continuing to aggressively invade new areas and colonize them [7]. Their hardiness, invasiveness, and colonizing ability have overcome all attempts so far to control or destroy them, irrespective of whether the attempts were based on chemical, biological, or mechanical methods. If some success has been achieved it has at best been local and temporary—often the weakening of the hold of one invasive species paving the way for another equally invasive species [3,8]. This oft-encountered inability to control the invasion and associated colonization of these weeds results in the production of billions of tonnes of phytomass across the world which has no utility value. Worse, this happens at the expense of soil nutrients and other natural resources which would otherwise have been used by diverse species or for agriculture.

There is another equally serious fall-out. Upon senescence, the phytomass of the weeds decays in the open—part aerobically and part anaerobically. Both processes generate global-warming gases, but the latter process is more harmful than the former because it leads to about 65% of the biodegrading organic carbon being converted to methane. As each molecule of methane has been estimated to have 25–34% greater global warming potential than that of carbon dioxide [9], the contribution to global warming of the latter is several times greater than the former.

Among the possible ways of utilizing the phytomass of invasive plants is vermicomposting. It has the special attribute that it can potentially lead to organic fertilizer of which almost limitless demand exists across the world. But past attempts to vermicompost lantana, parthenium, and ipomoea—indeed any other botanical species—have been unviable. The reasons have been elaborated recently [10–12] and essentially comprise of the inherent drawbacks of the conventional vermicomposting technology which necessitate pre-composting of the weeds and/or augmenting them with animal manure. These factors, together with the slow rate of the conventional vermireactors, make the vermicomposting of phytomass economically unviable. For similar reasons, past attempts in vermicomposting lantana, parthenium, and ipomoea—as summarized in Table 1—have not led to any viable process.

Table 1. A summary of past attempts at the utilization of lantana, parthenium, and ipomoea as feed in vermireactors. All vermireactors were operated in batch mode and no quantifiable measure has been given by any of the authors with which it was decided that vermicomposting had been completed.

Manner of the Weed Utilization; Reactor Size (If Stated)	Earthworm Species Employed	Duration after Which the Vermicompost Was Harvested	Main Findings	Reference
Fly ash was mixed with parthenium in different ratios in square pots of 30 cm × 30 cm × 30 cm	*Eisenia fetida*	Two-three months	Fly ash mixed with parthenium appeared to be a good feed for earthworm	[13]
Parthenium and cowdung were mixed in 1:2 ratio	*Perionyx excavatus*	Two-four months	Weeds can be used as a resource for making vermicompost	[14]
Parthenium was mixed with cowdung in circular plastic containers of 10 kg capacity	*E. fetida*	Three-and-a-half 31/2 months	Parthenium and cowdung in 1:3 ratio appeared optimum for the growth and reproduction of *E. fetida*	[15]
Ipomoea, cowdung and soil were mixed in earthen pots 5 kg capacity	*Eudrilus eugeniae*	Two months	Ipomoea can be converted into an 'environment-friendly' nutrient source	[16]
Parthenium was mixed with cowdung and loaded in cement tanks of 1 m depth	*E. eugeniae*	One-and-a-half months	Aromatics, aliphatics, alcohols, phenols, and polysaccharides are significantly decreased while nutritional levels are increased through vermicomposting	[17]
Cow dung, food industry sludge, water hyacinth and parthenium were mixed in a circular plastic tub loaded with 1 kg of the substrate.	*E. fetida*	Three months	Higher ratios of parthenium and water hyacinth resulted in higher vermiprocessing efficiency	[18]
Lantana was mixed with cowdung in different ratios.	*E. fetida*	Two months	Vermibeds with 40–60% of parthenium leaves showed better mineralization	[19]
Partheniumand cow dung mixtures were used incement tanks of 1 m depth.	*E. eugeniae*	The mixture was precomposted for 75 days and then harvesting of the vermicast was carried outonce in 15 days	Appropriate mixing of parthenium with cowdung is essential for the survival of the earthworms	[20]
Parthenium, farm wastes, and animal manure were mixed 10:1:1 in cement tanks of 1 m^3 volume.	*E. fetida*	Two months	Addition of different farm and animal wastes helped to degrade parthenium	[21]
Parthenium was mixed with biogas plant slurry in circular plastic tubs.	*E. fetida*	Two months	Parthenium mixed with biogas plant slurry could be 'profitably' vermicomposted	[22]

To overcome these hurdles the authors have developed the concept of 'high-rate vermicomposting' [10]. The authors have also designed and tested several machines aimed at translating the concept to application [23–26]. Further, as reported in an accompanying paper in this journal [27], the authors have found that upon being vermicomposted, lantana loses its toxicity and is transformed into an organic fertilizer as benign and potent as vermicompost derived from cowdung. However, this finding can be of practical utility only if technology is available to transform weeds such as lantana, parthenium, and ipomoea not only swiftly but also directly—i.e., without any pretreatment and without any fortification with animal manure. This report describes studies carried out with the objective of (a) developing such a process; (b) assessing the robustness and sustainability of the process when used uninterruptedly for several months; and (c) identifying factors, if any, with which process efficiency can be improved further.

2. Material and Method

2.1. Substrate and Vermicomposting

Leaves of each of the species were collected from their respective natural strands situated near the place of the author's work (Pondicherry University campus). They were rinsed with tap water to remove adhering muck and invertebrates—if any—and gently wiped before loading them into the HEVSTOW (high efficiency vertically stocked vermicomposting system for treating organic waste) vermicomposting machine described elsewhere [28]. HEVSTOW is a multi-module semi-continuous vermicomposting machine (Figure 1) designed on the basis of the high-rate vermicomposting concept reported earlier [10].

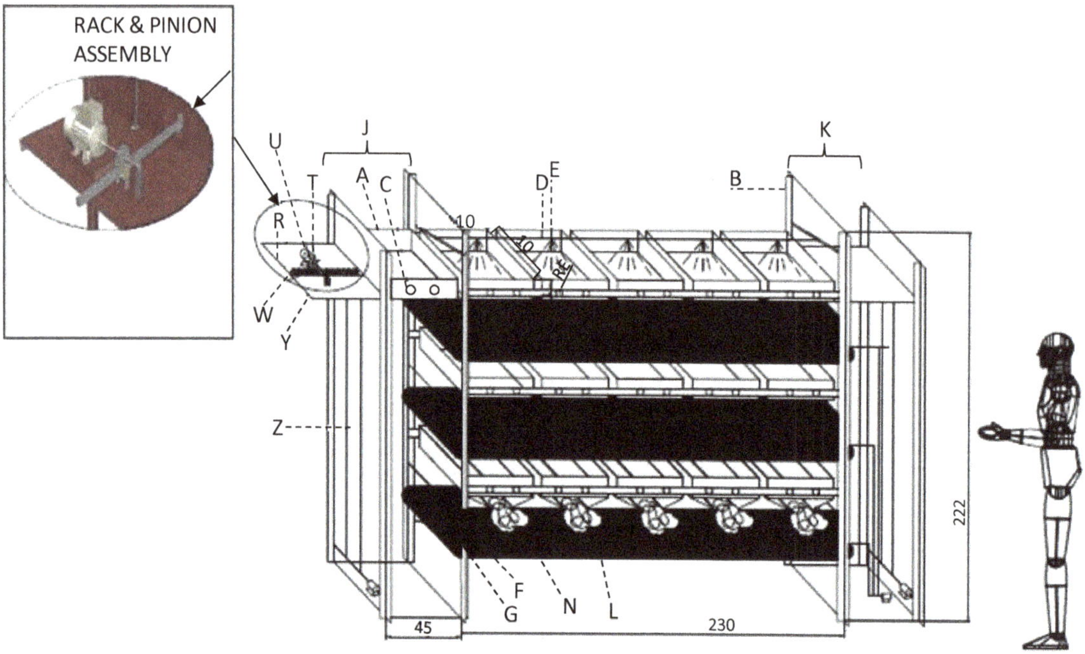

Figure 1. Schematic of the high efficiency vertically stacked vermicomposting system for treating organic waste (HEVSTOW); the human figure has been put to give an indication of the size.

HEVSTOW consists of a set of modular reactors and arrangements for their swift loading and unloading. A **fixed frame B** is provided to hold **modules A** loaded in series as well as in parallel. The modules move over B with the help of **wheels C** present on either side. The wheels are so positioned that they prevent A from moving vertically at the time of harvesting. A **sprinkler system D**, with **nozzles E** positioned above the modules, maintains the moisture content in A. There is a **rod F** placed below A, which can be rotated 180° using **gear mechanism G**. It helps in emptying the contents of the modules onto the **conveyor belt H** placed below each track. The guiding mechanism at one end of H enables the removal of the contents of the modules without any spillage. The **loading J** and **unloading K** systems help in the loading of A onto the fixed frame B or its unloading off B, using **rack and pinion arrangements R and T**. A roller attached to a motor helps in rotating H at the time of harvesting.

During operation, A is filled with substrate and placed on the loading end J. The motor aids in the lifting of the module with the help of a **rope Z**. The rack and pinion R and T arrangements, driven by **motor U**, help in placing module A onto the fixed frames B. In turn, U is supported on a frame and the **rod W** is attached to a **hinge X**. The whole set-up is placed on **frame Y**.

Each module in the HEVSTOW system used by us had 0.4 m × 0.4 m surface area and 0.12 m height. No chopping, pruning, soaking, or any other form of pre-treatment was performed. A jute cloth sheet of 3mm thickness, saturated with water, was provided at the bottom of each module to serve as bedding for the earthworms. The feed was laid over the jute cloth. The HEVSTOW prototypes used by these authors were fabricated from aluminum sheets of appropriate thickness, and steel bars/pipes. However, other appropriate materials such as fiberglass can be used in the manufacture of the HEVSTOW units.

In order to quantify the vermicast generation per adult worm, the modules were operated in the pseudo-discretized continuous reactor operation (PDCOP) mode, conceived by S. A. Abbasi and coworkers, and described elsewhere [29].

Its defining features are as summarized below:

- It enables reactor operation which is not actually continuous but approximates continuity; hence the term 'pseudo-discretized continuous'.
- In PDCOP, the vermireactors are initiated with a pre-set quantity of the substrate and a certain fixed number of adult earthworms. After allowing the earthworms to effect vermicomposting for a set number of days, say 20 or 25, the reactor contents are transferred to another container for determining the extent of conversion of the substrate to vermicast as also assessing the fecundity by counting the offspring in terms of the numbers of juveniles and cocoons produced by the earthworms. Soon after removing the reactor contents, the reactors are restarted with fresh weed feed but with the same adult animals that were deployed initially, while excluding the juveniles and cocoons. This makes it possible to measure the rate of vermicast production per adult animal and per unit of time.
- Since the unused substrate—which, if not removed, would have biodegraded even without the action of the earthworms—is removed every 20–25 days, the effect of factors other than ingestion of the feed by the earthworms is minimized.
- PDCOP thus ensures that the earthworms graze only upon totally fresh, or almost fresh, feed as they would be doing in the 'high-rate' vermireactor operation based on low solid retention times (SRTs) of just 20–25 days. Here SRT implies the time given in each pulse of feeding-harvesting for the earthworms to carry out vermicomposting. The lower the SRT needed for adequate vermicomposting, the higher the process efficiency. Further, since the juveniles and the cocoons that are generated in the vermireactors are separated before they could grow to the stage where they begin consuming significant quantities of the feed, their influence, too, on the reactor performance is sharply dampened.

All of the above enable assessment of the quality of vermicomposting garneted as a function of the number of earthworms and time, thereby providing avenues of process control and monitoring.

In the present work, three series of triplicate modules were started with 20, 50, or 80 earthworms for each weed, respectively, in the concerned modules. Each module was loaded with 1 kg dry weight equivalent of fresh weed. Healthy, adult, individuals of *E. fetida* were picked for this purpose randomly from cowdung-fed cultures maintained by the author. In the first run, all modules were allowed to function for 30 days after which they were emptied and their contents were transferred to separate containers for the assessment of vermicast and production of juveniles and cocoons. Immediately thereafter the reactors were started afresh in which everything else was kept the same as it was at the start of the experiment except that the adult earthworms removed from the previous run, were reintroduced into the fresh feed. Subsequent runs were of 20-day duration.

Throughout the experiments, all the modules were kept under identical ambient conditions of 30 °C ± 5 °C temperature and 60 ± 10% relative humidity. Their moisture level was maintained at 65 ± 5%. Mass balance of feed input and vermicast output was performed separately on the basis of respective dry weights taken after oven-drying their randomly picked and pooled samples at 105 °C to constant weight. To separate castings from other particles, the harvest was sieved through a 3 mm mesh.

2.2. *Physical and Chemical Characteristics*

Electrical conductivity (EC) and pH of the samples (vermicast and the parent weed) were measured in 1:2 (*v*/*w*) suspensions in water using EI™611E EC meter and Digison™ digital pH meter 7007, respectively. The bulk density and the particle density of the vermicast were measured on undisturbed cores by the graduated cylinder method [30] and the volumetric flask method [31], respectively. The vermicast's total porosity was then computed on the basis of its particle and bulk density values [30].

Total organic carbon was estimated using the modified dichromate redox method for the respective weeds and their vermicastas described by Heanes [32]. Total nitrogen was determined by the modified Kjeldahl method [33] for each vermicast and its parent weed using Kel Plus™ semi-automated digester and distillation units. The inorganic NH_4^+ and NO_3^- were determined by modified indophenol blue and Devarda's alloy methods, respectively [31,34] for vermicast and the weeds after they were extracted from the respective samples into a 2M KCl solution (1:10 *w*/*v*). Extractable/available potassium, calcium, and sodium were determined using Elico™ CL378 flame photometer after extraction from each vermicast or its parent substrate with Mehlich 3 extraction solution [35].Extractable/available copper, manganese, and zinc were determined using atomic emission spectroscopy (AES) by extracting the sample with Mehlich 3 extraction solution in a 1:25 sample-to-extractant ratio [35]. The same Mehlich 3 extract was used to determine the available phosphorus according to the ammonium molybdate–ascorbic acid method [36].

3. Results and Discussion

3.1. *Vermicomposting of Lantana*

3.1.1. Vermicast Production and Fecundity

The findings on the generation of vermicast and juveniles/cocoons produced during approximately 16 months of uninterrupted HEVSTOW operation in its 9 modules with 20, 50, and 80 adult individuals of *E. fetida* in triplicate sets are presented in Tables 2–4, respectively.

Due to logistics all modules could not be processed on the same day and had to be handled in a space of 2–3 days. As a result, the duration of the pulse varied by a day or two once in a while. Further, in modules with 80 earthworms, some harvests were carried out at 30 days intervals. However, since vermicast has been calculated in terms of per worm, per day, these variations do not cause any difficulty in comparing the observations across different reactors. Even though the vermicast production among triplicates varied from run to run (pulse to pulse), the overall average yield was in remarkably close agreement in all three sets. This reproducibility across triplicates extended to juveniles and cocoons as well, especially the former.

Table 2. Vermicomposting of lantana with 20 adults of *E. fetida* per kg of feed in pulse-fed modules.

Number of Days from the Start of the Reactor	Vermicast Generated per Worm (mg), per Day			Number of Juveniles Produced			Number of Cocoons Generated		
	Reactor I	Reactor II	Reactor III	Reactor I	Reactor II	Reactor III	Reactor I	Reactor II	Reactor III
30	12.4	32.1	26.3	0	1	4	7	5	4
50	32.2	27.0	28.2	0	0	2	8	12	14
70	19.2	16.3	33.3	6	6	9	5	7	8
90	21.9	21.0	20.5	6	2	4	12	14	9
110	48.4	43.2	41.0	4	7	6	11	9	4
130	39.2	27.0	29.5	2	7	3	14	11	12
150	20.9	22.9	18.6	6	10	4	4	6	7
172	31.0	38.1	36.3	4	2	0	6	3	8
192	26.9	25.7	28.5	4	7	6	6	3	4
212	37.9	34.8	36.0	6	7	8	4	5	4
232	42.5	34.4	38.5	7	10	9	10	16	12
254	28.6	27.5	35.9	4	2	3	7	9	6
276	32.2	38.2	29.0	3	5	6	2	7	3
296	52.3	62.1	44.2	4	6	2	5	3	7
317	50.8	52.2	48.1	3	5	2	6	4	3
337	50.7	48.8	51.0	0	2	2	0	3	4
360	49.2	42.1	44.4	0	2	1	2	1	2
380	52.5	43.3	53.2	2	1	2	0	2	1
400	41.2	31.0	41.0	3	2	4	2	1	3
422	40.9	50.3	37.0	3	1	2	2	5	4
444	32.5	45.9	51.2	3	2	2	4	1	3
464	34.1	37.7	33.3	3	2	0	2	1	2
485	47.2	45.6	54.4	3	4	3	2	5	4
Average ± SD	36.7 ± 11.7	36.8 ± 11.3	35.6 ± 12.1	3.3 ± 2.1	4.0 ± 3.0	3.7 ± 2.6	5.3 ± 3.8	5.8 ± 4.3	5.6 ± 3.5
Overall average		36.4 ± 11.5			3.7 ± 2.5			5.5 ± 3.8	
Average ± SD (of the last six month's data)	44.3 ± 7.5	44.1 ± 6.6	41.5 ± 15.8	2.2 ± 1.3	2.3 ±1.3	2.0 ± 1.1	2.2 ± 1.9	2.6 ± 1.7	2.9 ± 1.1
Overall average ± SD (of the last six month's data)		43.3 ± 10.4			2.2 ±1.2			2.6 ± 1.6	

Table 3. Vermicomposting of lantana with 50 adults of *E. fetida* per kg of feed in pulse-fed modules.

Number of Days from the Start of the Reactor	Vermicast Generated per Worm (mg), per Day			Number of Juveniles Produced			Number of Cocoons Generated		
	Reactor I	Reactor II	Reactor III	Reactor I	Reactor II	Reactor III	Reactor I	Reactor II	Reactor III
30	19.0	17.7	17.0	5	4	6	12	11	9
50	20.6	17.2	16.9	3	4	6	14	20	14
70	16.3	15.7	15.6	9	6	7	7	8	5
90	12.1	11.4	12.3	1	7	6	17	23	14
110	27.8	30.9	28.7	10	4	7	14	16	11
130	21.7	23.8	22.3	8	7	4	13	16	21
150	16.9	16.5	18.0	12	16	14	11	10	9
172	21.9	22.8	21.3	5	7	6	4	9	11
192	18.3	16.9	17.8	8	9	5	7	6	4
212	22.8	24.9	25.8	9	8	11	6	5	7
232	22.5	17.9	23.3	12	12	16	21	14	19
254	16.8	17.8	20.2	6	4	6	11	8	9
276	19.3	20.8	18.3	4	7	6	5	9	7
296	30.5	35.5	37.3	6	3	8	4	7	11
317	27.6	32.8	31.8	6	8	4	7	11	8
337	25.9	28.5	30.5	4	3	6	5	4	8
360	25.8	23.9	25.8	4	5	3	3	4	3
380	25.2	23.8	25.3	4	3	5	3	4	6
400	24.1	22.9	19.7	4	6	7	4	5	4
422	33.2	27.0	25.3	4	3	2	6	7	3
444	32.5	30.2	32.3	4	5	3	6	4	2
464	26.4	29.5	25.2	4	5	3	2	4	3
485	37.8	31.3	40.9	6	5	6	5	7	6
Average ± SD	23.7 ± 6.2	23.5 ± 6.4	24.0 ± 7.1	6 ± 2.9	6.1 ± 3.1	6.4 ±3.3	8.1 ± 5.0	9.2 ± 5.3	8.4 ± 5.0
Overall average		23.7 ± 6.5			6.2 ± 3.1			8.6 ± 5.1	
Average ± SD (of the last six month's data)	28.7 ± 4.1	27.8 ± 4.9	28.5 ± 13.1	4.4 ± 1.3	4.8 ± 1.7	4.3 ± 1.7	4.6 ± 2.1	5.6 ± 2.3	4.8 ± 2.3
Overall average ± SD (of the last six month's data)		28.3 ± 4.7			4.5 ± 1.4			5.0 ± 2.1	

Table 4. Vermicomposting of lantana with 80 adults of *E. fetida* per kg of feed in pulse-fed modules.

Number of Days from the Start of the Reactor	Vermicast Generated per Worm (mg), per Day			Number of Juveniles Produced			Number of Cocoons Generated		
	Reactor I	Reactor II	Reactor III	Reactor I	Reactor II	Reactor III	Reactor I	Reactor II	Reactor III
30	23.3	20.0	19.7	9	7	7	38	26	29
50	17.2	15.2	14.0	6	6	4	22	23	18
70	12.6	13.5	16.4	7	9	12	4	10	12
90	10.3	7.2	9.2	4	6	3	48	37	29
110	14.3	15.9	20.8	7	12	9	17	22	14
130	24.7	26.1	24.2	8	7	9	24	20	17
150	14.5	13.3	14.1	20	18	14	42	36	29
172	16.8	14.8	15.1	9	7	12	10	9	7
192	16.1	16.4	17.5	11	14	12	9	7	8
212	20.7	21.9	21.4	13	16	14	10	13	9
232	18.4	17.7	15.7	17	14	9	29	20	18
262	17.8	21.2	18.8	9	6	11	23	17	14
292	17.5	13.7	15.1	8	10	6	11	9	12
322	27.3	24.0	28.1	7	6	8	9	10	7
352	25.3	26.0	25.9	7	6	9	12	7	14
382	21.9	21.9	22.0	6	4	8	5	6	7
412	26.6	27.6	31.0	7	4	6	6	3	4
442	27.6	29.4	23.1	4	5	7	6	7	5
472	27.1	26.9	24.7	7	4	5	5	3	4
Average ± SD	20.0 ± 5.4	19.6 ± 6.1	19.8 ± 5.6	8.7 ± 4.1	8.5 ± 4.3	8.7 ± 3.2	17.4 ± 13.5	15.0 ± 10.3	13.5 ± 8.2
Overall average ± SD		19.8 ± 5.6			8.6 ± 3.8			15.3 ± 10.8	
Average ± SD (of the last six month's data)	24.8 ± 3.8	24.2 ± 5.2	24.3 ± 5.1	6.6 ± 1.3	5.6 ± 2.1	7.0 ± 1.4	7.7 ± 2.9	6.4 ± 2.7	7.6 ± 4.0
Overall average ± SD (of the last six month's data)		24.4 ± 4.5			6.4 ± 1.7			7.2 ± 3.1	

The trends in vermicast production as a function of duration for the three sets of reactors are presented in Figure 2a–c. The statistical trend lines show a rising trend in all three cases, indicating that with time the earthworms—which had been born and grown in cowdung-fed cultures—increasingly adapted to the lantana feed. Indeed, the average vermicast output during the last six months of the experiment was substantially higher than the overall average (Tables 2–4). It also indicates that more prolonged reactor operation as also the use of *E. fetida* offspring, who are born and grown in lantana-fed cultures, are likely to yield higher vermicast per animal than the maximum achieved in our experiments.

As expected, the modules which had just 20 earthworms per kg (fresh weight) of lantana generated the maximum vermicast per worm (per day) due to the most liberal availability of the feed and hence the easiest access to it of the three module types. In modules with $2\frac{1}{2}$ times this population, viz 50 earthworms per kg of lantana, competition for access to food brought the per capita yield down (Table 3). In still more crowded reactors operated with 80 earthworms per kg of lantana, the per capita vermicast production was still lower (Table 3), but the margin of difference was not as pronounced as it was between reactors with 20 earthworms and 50 earthworms per kg of lantana.

In terms of absolute vermicast production, and if the average of the last six month's data is used as the base—which is logical, given that due to the rising trend future yields are likely to be at least as good, possibly better—the situation is as explained below lantana had 22.4% dry weight. Hence, each kg fresh weight of lantana contained 224g of solids.

The modules with 20 earthworms per kg fresh weight (or 224 g dry weight) of lantana generated $(43.3 \times 20 \times 30)/1000 = 25.98$ (rounded to 26) g of vermicast (dry weight equivalent) per month. In other words, converting 11.6% of the feed to vermicast per month.

The modules with 50 earthworms per kg fresh weight (or 224 g dry weight) of lantana generated $(28.3 \times 50 \times 30)/1000 = 42.45$ (rounded to 42.5) g of vermicast (dry weight equivalent) per month. In other words, converting 19% of the feed to vermicast per month.

The modules with 80 earthworms per kg fresh weight (or 224 g dry weight) of lantana generated $(24.4 \times 80 \times 30)/1000 = 58.56$ (rounded to 58.6) g of vermicast (dry weight equivalent) per month. In other words, converting 26.2% of the feed to vermicast per month.

Given that 50 ± 10% of organic carbon contained in any feed is either converted to worm zoomass or is lost as CO_2(due to respiration by earthworms and microorganisms present in the feed) in the course of vermicomposting, the above-mentioned figures reflect the conversion of about twice as much feed as the vermicast produced. Hence the effective conversion of feed to vermicast per month in reactors with 80 earthworms is equivalent to 52.4 ± 10% utilization of the feed per month. But the rising trend in vermicast production with time (Figure 2a–c) means vermicast output is set to increase further with time. Secondly, had we not been removing the juveniles and cocoons from the modules, they would be utilizing substantial parts of the feed. The combination of both these factors is likely to have caused much more than 52.4 ± 10% utilization of lantana per month and the actual vermicast yield would have approached its theoretical maximum at 30–40-day SRTs. This rate is several times faster than the 90–120 days that are taken by conventional vermireactors. Equally important, this rate has been achieved without any pre-composting, cowdung supplementation, or even any pre-treatment of the lantana feed.

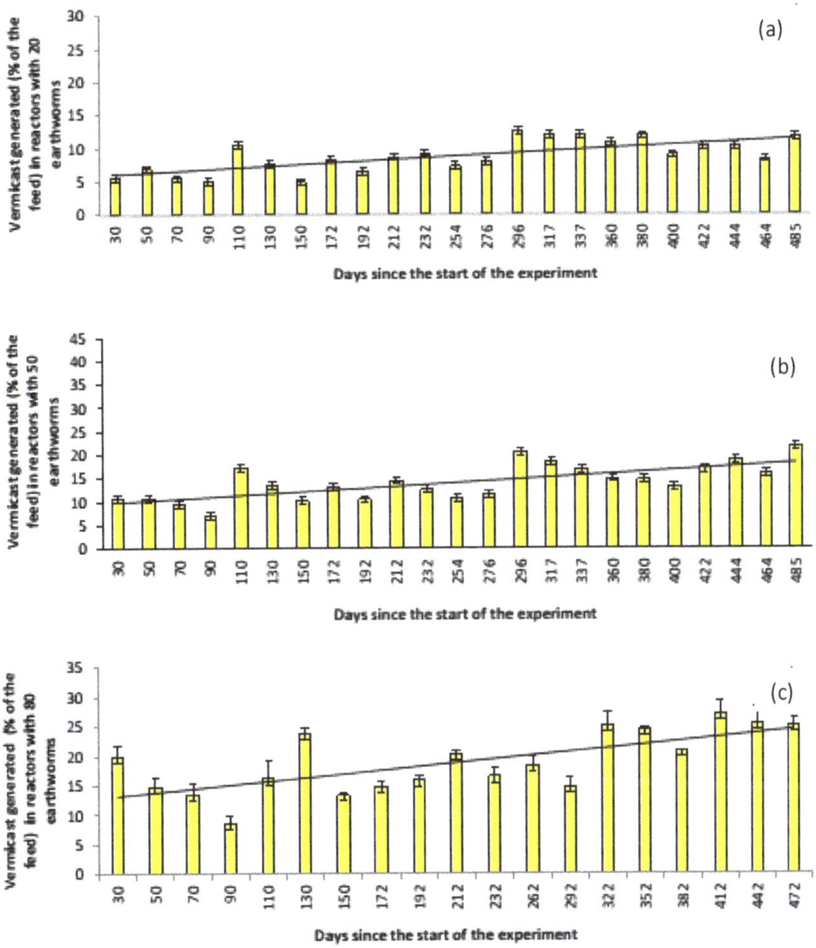

Figure 2. Trend in the generation of vermicast as a function of time in pulse-fed, semi-continuous reactors operated with (**a**) 20, (**b**) 50, and (**c**) 80 earthworms and fed with fresh lantana.

Depending on species and variety, individual earthworms take 6–12h for converting the material they ingest into their vermicast [37]. If a means can be found to immobilize live earthworms in a way that each can be fed individually and its cast harvested, as soon as it is exited, vermicomposting of any substrate should not take more than 6–12h. However, it is not possible to engineer vermicomposting systems which can accomplish this. In a vermireactor each earthworm has to first find food in competition with other earthworms. It then has to leave its casting in the midst of the feed, making its immediate and clean-cut harvesting almost impossible. As a result, the product of vermicomposting becomes fit for harvesting only when a large fraction of the parent substrate has been converted to vermicast. In conventional vermireactors this becomes possible after 90–120 days. The paradigm shift achieved in high-rate vermicomposting shortens this duration to 20–30 days but further improvements in increasing the rate of vermicomposting appear unlikely. This is due to the engineering limits associated with maximizing access to food and speeding up the harvesting of the vermicast.

3.1.2. Chemical Characteristics of the Lantana Vermicompost in Comparison to Lantana

Vermicomposting of lantana is seen to have caused significant differences to arise between the vermicast and its parent substrate (Table 5). The total organic carbon (TOC) content, which was 453.6 g/kg in lantana falls to 248.7 g/kg in the weed's vermicompost, reflecting a 57.4% reduction. Concurrently, there is an increase in total nitrogen from 16 to 18 g/kg, calculated on the basis of initial feed mass, perhaps by way of mucus contributed by the earthworms. The combination of these two factors causes a reduction in the carbon-to-nitrogen (C:N) ratio of the vermicast relative to lantana—from 28 to 14. This plays a major role in making lantana vermicast a highly potent fertilizer because a C/N ratio of less than 20 in an organic fertilizer makes it acceptable for use while a C:N ratio of 15 or less is deemed ideal [38,39]. Vermicomposting thus transforms lantana into a nitrogen-rich fertilizer of the ideal C:N ratio.

Table 5. Chemical characteristics of lantana and its vermicast.

Variables	Values in	
	Lantana	Vermicast
Total organic carbon (g/kg)	453.6 ± 12.5	248.7 ± 5
Total nitrogen (g/kg)	16 ± 0.4	18 ± 1.1
C:N ratio	28:1	14:1
Ammoniacal nitrogen (mg/kg)	-	321 ± 5.4
Nitrate nitrogen (g/kg)	1.73 ± 0.05	14 ± 0.5
Available sodium (g/kg)	0.080 ± 0.010	0.260 ± 0.0051
Available potassium (g/kg)	1.023 ± 0.012	4.1 ± 0.2
Available calcium (g/kg)	1.08 ± 0.09	4.3 ± 0.23
Available phosphorous (mg/kg)	79.8 ±2.1	324.38 ± 20.1
Total copper (mg/kg)	21.27 ± 1.86	32.1 ± 7.05
Available copper (mg/kg)	3.33 ± 1.155	13.53 ± 0.73
Total manganese (mg/kg)	128.8 ± 9.73	163.9 ± 18.51
Available manganese (mg/kg)	10.3 ± 2.32	87.9 ± 0.3
Total zinc (mg/kg)	106.33 ± 8.42	123.1 ± 35.18
Available zinc (mg/kg)	18.33 ± 1.33	58.56 ± 0.58

There is an 8-fold increase in nitrate nitrogen reflecting the high degree of mineralization occurring when lantana is transformed into vermicast. There is an equally dramatic increase in available sodium, potassium, calcium, and phosphorous in vermicast relative to lantana. The levels of total copper, total manganese, and total zinc have also increased mildly, while those of available copper, manganese, and zinc have gone up dramatically.

All these characteristics point towards lantana having been converted by vermicomposting into a potential fertilizer.

3.2. Vermicomposting of Parthenium

3.2.1. Vermicast Production and Fecundity

All modules had vermicast production steadily rising with time as seen in the trend lines (Figure 3a–c). There was steady production of juveniles and cocoons in all the reactors.

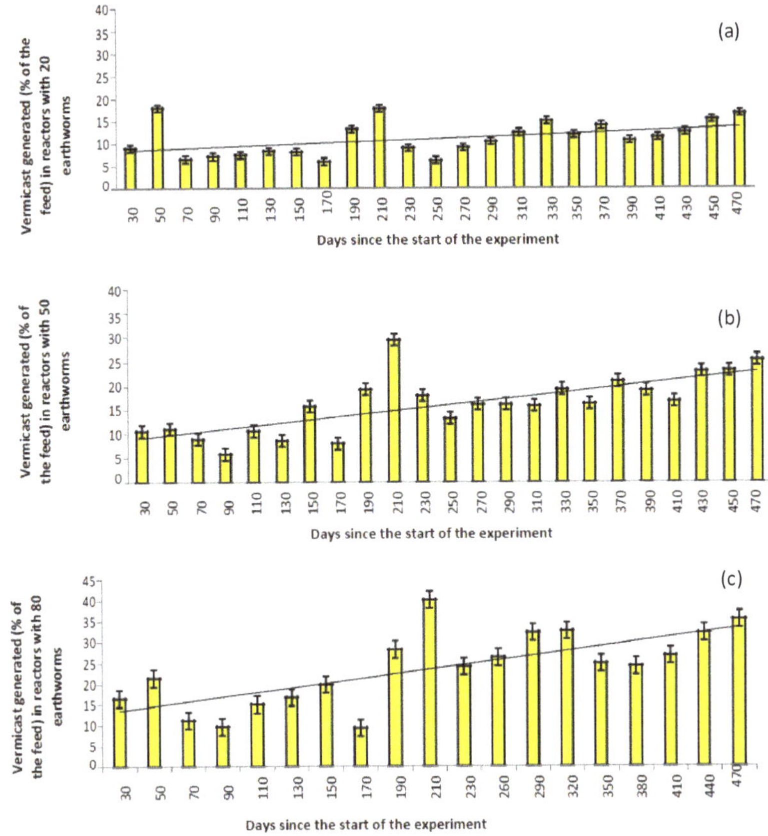

Figure 3. Trend in the generation of vermicast as a function of time in pulse-fed, semi-continuous reactors operated with (**a**) 20, (**b**)50, and (**c**) 80 earthworms with fresh Parthenium.

If figures of average vermicast production per earthworm during the last six months of the system operation per day are used to calculate the fraction of parthenium converted to vermicast per month, in the same manner as illustrated with lantana in Section 3.1.1, the corresponding figures are as follows.

In reactors with 20 earthworms per kg (equivalent to 305 g dry weight) of parthenium, the vermicast generated per month is 7.5% of the feed mass. In reactors with 50 and 80 earthworms per kg (equivalent to 305 g dry weight) of parthenium, the vermicast generated per month is 11.3% and 17.5% of the feed mass. Considering that (a) with time there is increasing adaptation of earthworms to parthenium feed as also to the confines of the HEVSTOW modules; (b) the juveniles and cocoons if not removed from the system would have contributed to even greater utilization of the feed, and (c) the effective utilization of feed is about twice as much as the vermicast produced (due to the loss of about half of the feed in metabolism), it can be safely assumed that with time the rate of parthenium utilization would significantly improve in HEVSTOW to achieve near total conversion to vermicast in 30–40 days.

The results are summarized in Tables 6–8. In the case of Parthenium-fed modules also, the averages of the vermicast yield in the triplicates were in close agreement even as the output of constituent runs varied. The per worm output of vermicast in modules with 20 earthworms per kg (equivalent to 305 g dry weight) of parthenium was significantly higher than the per animal output in reactors with 50 earthworms per kg of parthenium, evidently due to the liberal availability of the feed in the former case. However, a further increase in earthworm density to 80 animals per kg (Table 8) did not cause any significant change in per capita vermicast production. The greater crowding did seem to affect the rate of vermicomposting in the initial months due to which the overall average vermicast output in reactors with 50 earthworms—18.8 ± 7.4 mg/worm/day—is higher than the overall average—17.6 ± 6.9 mg/worm day in reactors with 80 earthworms. However, this difference has disappeared during the last six months of the system operation and the average output during the last six months in the two types of modules is almost the same. This indicates a possible adaptation with time not only with parthenium as the sole feed but also with the higher earthworm density. It also indicates that the overall vermicast production in reactors with 80 earthworm/kg will be much higher than in reactors with 50 earthworms/kg because the per capita vermicast production in the reactors of these two animal densities become close to each other once the adaptation to higher animal density is over.

Table 6. Vermicomposting of parthenium with 20 adults of *E. fetida* per kg of feed in pulse-fed modules.

Number of Days from the Start of the Reactor	Vermicast Generated per Worm (mg), per Day			Number of Juveniles Produced			Number of Cocoons Generated		
	Reactor I	Reactor II	Reactor III	Reactor I	Reactor II	Reactor III	Reactor I	Reactor II	Reactor III
30	19.9	35.8	22.7	1	3	0	2	3	3
50	55.4	54.2	48.3	3	0	4	0	2	2
70	20.8	18.1	18.5	2	6	4	4	3	5
90	23.5	17.1	21.5	0	3	0	4	6	3
110	19.2	23.5	22.1	2	0	1	3	4	3
130	20.0	27.1	24.8	0	0	0	0	0	0
150	22.1	24.2	24.4	4	3	6	0	0	0
170	18.1	15.4	18.5	4	0	2	2	1	2
190	34.2	39.2	42.1	3	5	3	4	2	2
210	53.3	42.1	60.2	3	6	2	4	5	5
230	18.1	24.0	36.0	0	2	0	2	0	0
250	12.1	25.4	17.1	2	0	0	1	2	0
270	24.2	18.3	36.5	0	2	1	0	1	0
290	38.8	25.2	27.1	0	3	2	1	2	4
310	36.7	41.3	31.0	0	2	2	0	0	1
330	46.0	46.5	38.8	0	0	2	0	2	3
350	34.4	35.0	35.6	0	0	0	2	0	3
370	40.4	42.5	39.4	0	1	0	0	2	0
390	37.3	31.9	24.6	0	0	0	1	0	2
410	35.4	32.1	32.3	0	1	2	0	2	1
430	39.4	29.4	40.8	0	2	0	0	1	0
450	49.2	43.1	41.9	0	0	2	0	0	1
470	47.3	46.5	52.5	2	3	3	2	2	3
Average ± SD	32.4 ± 12.7	32.1 ± 10.9	32.9 ± 11.6	1.1 ± 1.5	1.8 ± 1.9	1.6 ± 1.6	1.4 ± 1.5	1.7 ± 1.7	1.9 ± 1.6
Overall average ± SD		32.5 ± 11.6			1.5 ± 1.7			1.7 ± 1.6	
Average ± SD (of the last six month's data)	40.5 ± 5.2	37.3 ± 7.6	36.4 ± 8.1	0.2 ± 0.6	1.2 ± 1.2	1.3 ± 1.2	0.6 ± 0.8	1.1 ± 1.0	1.8 ± 1.4
Overall average ± SD (of the last six month's data)		38.1 ± 7.1			0.9 ± 1.1			1.2 ± 1.2	

Table 7. Vermicomposting of parthenium with 50 adults of *E. fetida* per kg of feed in pulse-fed modules.

Number of Days from the Start of the Reactor	Vermicast Generated per Worm (mg), per Day			Number of Juveniles Produced			Number of Cocoons Generated		
	Reactor I	Reactor II	Reactor III	Reactor I	Reactor II	Reactor III	Reactor I	Reactor II	Reactor III
30	10.0	9.7	16.7	5	6	6	5	4	7
50	13.7	7.3	17.7	6	7	5	4	6	3
70	9.3	9.8	11.8	1	7	2	6	4	2
90	6.7	7.7	5.8	6	4	2	4	3	3
110	10.0	15.9	11.2	4	6	4	7	9	3
130	10.8	9.3	9.8	0	0	0	0	0	0
150	19.3	17.7	18.0	11	10	7	4	8	5
170	9.2	9.4	9.3	5	4	6	4	3	2
190	22.3	21.4	23.8	6	4	7	4	5	3
210	34.7	34.0	33.8	5	3	7	4	3	6
230	23.8	17.8	21.5	2	3	2	1	0	2
250	12.8	15.4	18.1	5	4	1	2	0	3
270	16.5	21.3	18.7	2	3	0	0	2	1
290	20.8	14.4	21.3	4	7	3	3	4	5
310	20.3	19.3	15.8	3	4	2	3	1	4
330	22.6	24.8	20.3	2	4	4	3	5	2
350	20.9	18.3	17.8	2	1	4	3	2	4
370	29.4	25.6	18.2	0	3	2	2	4	1
390	23.1	27.2	16.7	2	1	0	3	2	2
410	19.8	26.1	12.9	4	0	3	2	3	0
430	30.2	28.2	22.3	2	3	0	2	4	2
450	28.0	26.7	26.3	3	4	2	2	3	1
470	31.3	28.4	29.2	4	5	4	4	3	2
Average ± SD	19.4 ± 8.0	18.9 ± 7.8	18.1 ± 6.5	3.7 ± 2.4	4.0 ± 2.4	3.2 ± 2.3	3.1 ± 1.7	3.4 ± 2.3	2.7 ± 1.8
Overall average ± SD		18.8 ± 7.4			3.6 ± 2.4			3.1 ± 1.9	
Average ± SD (of the last six month's data)	24.6 ± 4.6	23.9 ± 4.8	20.1 ± 4.9	2.6 ± 1.3	3.2 ± 2.1	2.4 ± 1.5	2.7 ± 0.7	3.1 ± 1.2	2.3 ± 1.6
Overall average ± SD (of the last six month's data)		22.9 ± 5.0			2.7 ± 1.6			2.7 ± 1.2	

Table 8. Vermicomposting of parthenium with 80 adults of *E. fetida* per kg of feed in pulse-fed modules.

Number of Days from the Start of the Reactor	Vermicast Generated per Worm (mg), per Day			Number of Juveniles Produced			Number of Cocoons Generated		
	Reactor I	Reactor II	Reactor III	Reactor I	Reactor II	Reactor III	Reactor I	Reactor II	Reactor III
30	13.3	11.0	11.7	11	7	8	8	14	7
50	7.8	20.4	18.7	12	11	17	4	7	6
70	9.8	6.5	8.0	7	6	8	6	4	5
90	8.2	7.8	5.0	6	4	2	2	1	4
110	11.4	11.0	10.4	3	0	2	4	2	3
130	13.5	10.1	12.8	0	0	0	0	0	0
150	13.1	14.8	15.3	14	9	16	7	8	4
170	6.7	6.5	6.9	7	6	5	4	3	5
190	20.3	20.9	20.5	8	7	6	6	5	5
210	33.5	31.7	22.5	6	4	7	7	12	11
230	17.3	14.5	20.7	3	3	4	5	2	1
260	17.7	20.7	19.1	2	6	4	3	4	7
290	22.1	23.2	25.3	6	3	5	3	0	6
320	25.6	25.5	20.2	3	6	7	2	4	3
350	18.9	18.3	17.1	3	3	2	2	1	0
380	18.0	17.5	17.3	4	5	2	3	2	6
410	22.2	17.0	19.2	2	3	0	4	3	2
440	24.8	22.8	22.8	2	1	2	2	4	3
470	25.1	25.5	26.9	2	1	3	3	2	1
510	28.4	26.7	26.5	8	6	4	5	4	6
Average ± SD	17.9 ± 7.4	17.6 ± 7.2	17.3 ± 6.4	5.5 ± 3.8	4.6 ± 2.9	5.2 ± 4.5	4.0 ± 2.1	4.1 ± 3.7	4.3 ± 2.7
Overall average ± SD		17.6 ± 6.9			5.1 ± 3.8			4.1 ± 2.9	
Average ± SD (of the last six month's data)	23.3 ± 3.8	21.9 ± 4.2	21.4 ± 4.1	3.4 ± 2.1	3.6 ± 2.1	2.9 ± 2.2	3.0 ± 1.2	2.9 ± 1.2	3.0 ± 2.3
Overall average ± SD (of the last six month's data)		22.2 ± 3.9			3.3 ± 2.1			3.0 ± 1.6	

3.2.2. Chemical Characteristics of the Vermicast Relative to the Substrate

Upon vermicomposting parthenium loses about 25% of its TOC, leading to a change in the C:N ratio from 18 to 12. There is extensive mineralization, evidenced by the increase in nitrate nitrogen, and in the levels of available phosphorous, sodium, potassium, calcium, copper, manganese, and zinc levels (Table 9). In most cases, the increase is of several orders of magnitude (as in the case of available phosphorous, copper, manganese, and zinc). These changes, together with the fall in the C:N ratio below 15, indicate that parthenium has potentially turned into a fertilizer.

Table 9. Chemical characteristics of parthenium and its vermicast.

Variables	Values in	
	Parthenium	Vermicast
Total organic carbon (g/kg)	312 ± 7	234 ± 13
Total nitrogen (g/kg)	17 ± 0.2	20 ± 1.5
C:N ratio	18:1	12:1
Ammoniacal nitrogen (mg/kg)	-	262.5 ± 6.9
Nitrate nitrogen (g/kg)	1.31 ± 0.082	16.2 ± 1.4
Available sodium (g/kg)	0.145 ± 0.012	0.326 ± 0.0013
Available potassium (g/kg)	1.142 ± 0.015	2.5 ± 0.1
Available calcium (g/kg)	1.15 ± 0.11	3.2 ± 0.6
Available phosphorous (mg/kg)	116.4 ± 3.1	402.7 ± 5.6
Total copper (mg/kg)	24.9 ± 1.47	35.6 ± 11.05
Available copper (mg/kg)	0.37 ± 0.152	7.8 ± 0.23
Total manganese (mg/kg)	70.33 ± 16.62	88.6 ± 26.81
Available manganese (mg/kg)	7 ± 2.2	69.9 ± 1.2
Total zinc (mg/kg)	148.27 ± 9.32	173.2 ± 6.29
Available zinc (mg/kg)	3.9 ± 0.503	44.96 ± 2.27

3.3. *Vermicomposting of Ipomoea*

3.3.1. Vermicast Production and Fecundity

The findings are summarized in Tables 10–12. In terms of reproducibility of average output in triplicates—even as data of individual runs fluctuated from module to module—ipomoea-fed modules behaved in the same manner as the modules fed with lantana and parthenium. However, ipomoea-fed systems significantly deferred from these of the other two weeds in that the average output during the last six months did not vary substantially from the average output of the earlier months. Thus, earthworms seem to have adapted to the ipomoea feed straightaway. Accordingly, the statistical trend lines were more or less flat (Figure 4a–c).

Ipomoea also differed from other feeds in the sense that crowding of earthworms seemed to effect the per capita vermicast generation more than it did for the other two feeds, as reflected in an almost 50% drop in 50 animals per kg reactors compared to the 20 animals per kg reactors.

Following the methodology of converting the average per capita vermicast production of the last six months of the experiment to percent utilization of feed per month, we see that in modules with 20 earthworms per kg (or 221 g dry weight equivalent) of ipomoea, the vermicast generated is 8.9% of the feed. In modules with 50 and 80 earthworms, the corresponding figures are 13.8% and 19.5%, respectively. With higher earthworm density and by retaining the juveniles and cocoons in the modules the utilization per month for vermicast production can be taken to 50% or higher, thereby attaining full utilization in about 60 days. This rate is still significantly faster than the period of 90–120 days needed by conventional vermireactors which also require liberal supplementation of cowdung (in 1:1 or higher manure-ipomoea ratios) to utilize half of the same quantity of ipomoea.

Table 10. Vermicomposting of ipomoea with 20 adults of *E. fetida* per kg of feed in pulse-fed modules.

Number of Days from the Start of the Reactor	Vermicast Generated per Worm (mg), per Day			Number of Juveniles Produced			Number of Cocoons Generated		
	Reactor I	Reactor II	Reactor III	Reactor I	Reactor II	Reactor III	Reactor I	Reactor II	Reactor III
30	44.8	41.4	32.6	7	7	6	3	4	1
50	19.8	24.2	22.7	7	4	3	6	12	11
70	17.4	21.4	20.0	5	4	7	2	3	2
90	39.1	34.9	40.0	2	4	4	3	6	2
110	29.4	35.0	26.0	4	3	6	2	4	6
130	31.3	26.5	29.0	5	4	6	4	3	3
150	45.4	40.6	36.0	6	4	6	5	7	7
173	28.3	21.5	36.3	5	3	0	3	4	2
193	32.9	29.0	40.1	2	1	3	4	0	6
213	24.3	29.3	35.0	2	0	3	1	3	2
233	31.0	38.2	33.3	6	4	7	2	1	3
253	22.0	27.3	27.0	4	3	6	2	0	3
273	19.0	23.8	20.0	2	1	4	1	0	2
293	20.9	21.4	21.4	3	2	4	2	0	3
313	29.4	28.3	21.1	3	1	2	2	4	1
333	35.1	23.2	33.2	2	1	3	1	4	2
353	44.9	42.9	34.9	2	1	2	0	4	3
373	44.5	39.5	41.7	2	1	4	2	3	2
393	43.6	41.5	35.3	2	1	2	1	2	3
413	47.1	44.8	46.5	2	1	3	2	3	4
Average ± SD	32.5 ± 10.1	31.7 ± 8.2	31.6 ± 7.9	3.7 ± 1.8	2.5 ± 1.8	4.1 ± 1.9	2.4 ± 1.5	3.4 ± 2.8	3.4 ± 2.4
Overall average ± SD		32.0 ± 8.6			3.4 ± 1.9			3.1 ± 2.3	
Average ± SD (of the last six month's data)	33.8 ± 10.9	33.1 ± 9.1	31.4 ± 9.0	2.8 ± 1.3	1.6 ± 1.1	3.7 ± 1.7	1.5 ± 0.7	2.1 ± 1.7	2.6 ± 0.8
Overall average ± SD (of the last six month's data)		32.8 ± 9.4			2.7 ± 1.6			2.1 ± 1.2	

Table 11. Vermicomposting of ipomoea with 50 adults of *E. fetida* per kg of feed in pulse-fed modules.

Number of Days from the Start of the Reactor	Vermicast Generated per Worm (mg), per Day			Number of Juveniles Produced			Number of Cocoons Generated		
	Reactor I	Reactor II	Reactor III	Reactor I	Reactor II	Reactor III	Reactor I	Reactor II	Reactor III
30	19.4	18.8	16.3	10	12	9	6	7	10
50	11.5	9.9	10.6	12	10	9	14	17	22
70	13.8	13.6	15.0	7	5	9	3	4	8
90	24.9	29.0	29.7	5	4	7	3	6	5
110	24.5	21.3	22.7	6	7	5	5	9	7
130	18.6	20.0	22.3	7	8	7	5	6	4
150	26.6	30.4	27.7	10	8	14	8	9	11
173	31.8	20.5	28.4	7	6	5	6	4	3
193	23.5	27.8	26.9	4	6	2	5	7	3
213	23.3	29.7	24.2	3	0	5	4	2	6
233	29.6	31.3	33.9	11	7	9	7	5	4
253	20.9	20.5	23.2	7	6	5	4	3	7
273	16.2	19.0	16.5	4	6	3	3	4	5
293	13.0	13.2	15.6	5	6	4	3	5	4
313	13.0	14.1	14.0	3	6	5	4	3	6
333	16.0	14.4	14.8	4	3	3	4	5	2
353	20.4	22.1	31.9	3	6	4	4	5	3
373	20.4	22.6	26.5	3	6	3	6	4	5
393	19.7	19.8	19.7	3	4	5	5	4	6
413	22.5	23.0	23.5	3	4	6	6	5	7
Average ± SD	20.5 ± 5.6	21.1 ± 6.2	22.2 ± 6.6	5.9 ± 2.9	6.0 ± 2.5	6.0 ± 2.9	5.3 ± 2.5	5.7 ± 3.2	6.4 ± 4.3
Overall average ± SD		21.2 ± 6.1			5.9 ± 2.7			5.8 ± 3.4	
Average ± SD (of the last six month's data)	19.2 ± 5.0	20.0 ± 5.4	22.0 ± 7.1	4.6 ± 2.6	5.4 ± 1.3	4.7 ± 1.8	4.6 ± 1.3	4.3 ± 0.8	4.9 ± 1.7
Overall average ± SD (of the last six month's data)		20.4 ± 5.8			4.9 ± 1.9			4.6 ± 1.3	

Table 12. Vermicomposting of ipomoea with 80 adults of *E. fetida* per kg of feed in pulse-fed modules.

Number of Days from the Start of the Reactor	Vermicast Generated per Worm (mg), per Day			Number of Juveniles Produced			Number of Cocoons Generated		
	Reactor I	Reactor II	Reactor III	Reactor I	Reactor II	Reactor III	Reactor I	Reactor II	Reactor III
30	9.9	11.8	10.6	14	16	11	12	14	16
50	11.8	11.0	11.2	17	14	21	12	16	7
70	14.9	12.4	12.3	9	7	8	12	14	11
90	27.0	23.9	23.2	8	13	11	12	7	9
110	21.7	20.2	22.3	8	6	9	9	11	10
130	18.2	21.2	17.4	12	14	12	10	11	9
150	23.9	24.8	20.6	17	12	9	12	11	8
173	33.1	28.0	25.9	9	12	16	11	7	5
203	19.5	20.6	22.3	7	5	4	2	6	8
233	20.5	20.5	18.5	6	4	7	2	3	5
265	19.3	17.0	19.3	4	6	5	3	2	4
295	15.5	17.7	15.5	4	6	5	3	4	7
325	14.2	16.3	15.2	5	4	7	6	3	4
355	16.7	16.8	13.9	5	4	6	6	7	6
388	17.3	11.0	17.5	5	6	4	4	6	7
418	23.1	25.2	24.9	6	8	5	7	4	6
Average ± SD	19.2 ± 5.8	18.7 ± 5.3	18.2 ± 4.8	8.5 ± 4.3	8.6 ± 4.2	8.8 ± 4.7	7.7 ± 4.0	7.9 ± 4.4	7.6 ± 3.0
Overall average ± SD		18.7 ± 5.2			8.6 ± 4.3			7.7 ± 3.8	
Average ± SD (of the last six month's data)	18.1 ± 3.1	17.8 ± 4.3	17.8 ± 3.7	5.0 ± 0.8	5.4 ± 1.5	5.6 ± 1.1	4.4 ± 1.9	4.1 ± 1.8	5.6 ± 1.3
Overall average ± SD (of the last six month's data)		17.9 ± 3.5			5.3 ± 1.2			4.7 ± 1.7	

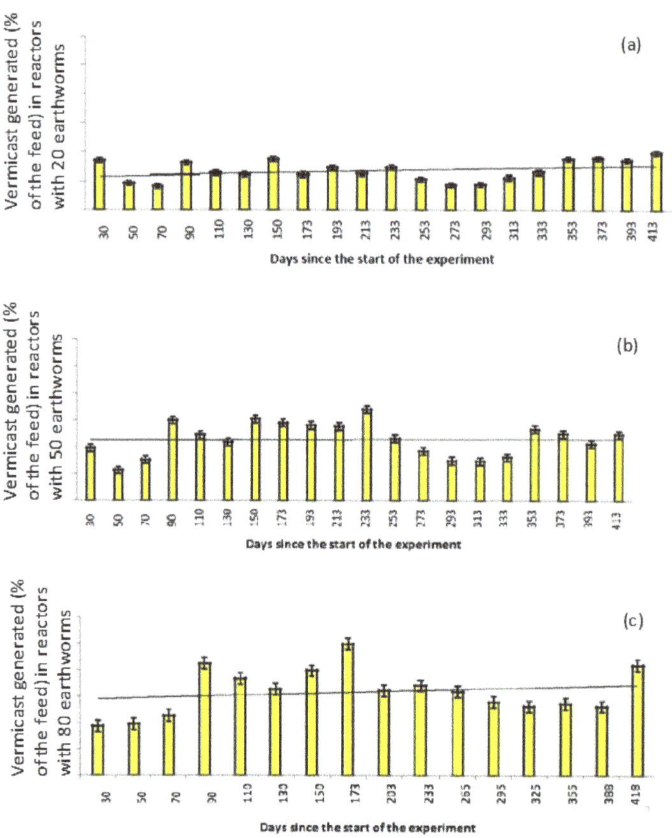

Figure 4. Trend in the generation of vermicast as a function of time in pulse-fed, semi-continuous reactors operated with (**a**) 20, (**b**) 50, and (**c**) 80 earthworms with fresh ipomoea.

3.3.2. Chemical Characteristics of Ipomoea Vermicast Relative to the Parent Substrate

Ipomoea loses 53% of its TOC in the process of getting converted to vermicast (Table 13) and its C:N ratio falls from 21 to 10,which is a level highly desirable in an organic fertilizer. It also gets extensively mineralized by having its nitrate nitrogen, available phosphorous, and available sodium, potassium, calcium, copper, manganese, and zinc increased in concentration by several orders of magnitude.

Table 13. Chemical characteristics of ipomoea and its vermicast.

Variables	Values in	
	Ipomoea	**Vermicast**
Total organic carbon (g/kg)	438 ± 15.3	233.3 ± 16.6
Total nitrogen (g/kg)	21 ± 0.7	23 ± 0.8
C:N ratio	21:1	10:1
Ammoniacal nitrogen (mg/kg)	-	237 ± 8.3
Nitrate nitrogen (g/kg)	1.8 ± 0.06	15.4 ± 1.5
Available sodium (g/kg)	0.072 ± 0.006	0.246 ± 0.0042
Available potassium (g/kg)	1.048 ± 0.012	3.6 ± 0.1
Available calcium (g/kg)	1.24 ± 0.18	4.5 ± 0.1
Available phosphorous (mg/kg)	89.9 ± 4.7	478 ± 6.2
Total copper (mg/kg)	40.2 ± 4.65	53 ± 6.08
Available copper (mg/kg)	1.33 ± 0.99	18.73 ± 0.37
Total manganese (mg/kg)	142.67 ± 7.02	183.2 ± 17.75
Available manganese (mg/kg)	17.3 ± 1.86	91.2 ± 1.7
Total zinc (mg/kg)	152.2 ± 7.53	181.1 ± 17.32
Available zinc (mg/kg)	13.8 ± 1.6	88.3 ± 1.4

3.4. *Fertilizer Value of the Vermicasts*

While these studies were being carried out, another group in the author's laboratory was parallelly investigating the fertilizer value of the vermicast of lantana, parthenium, ipomoea, salvinia, and prosopis (*Prosopis juliflora*). It carried out studies on germination and early growth [4,40–43] as well as full plant life up to the end of the fruit yield [44–47] of several vegetables with or without fertilization by these weed's vermicomposts. The studies showed that the vermicomposts of all the weeds were as plant friendly and soil-friendly as manure-based vermicasts are known to be [48]. The group also explored the causes behind the transformation of the toxic weeds into benign fertilizers [49–52]. It was seen that a) the chemicals responsible for the toxicity and allelopathy of these weeds were destroyed in the course of vermicomposting, and b) there was mineralization in the form of degradation of organic carbon into CO_2 (which escaped into the atmosphere) and of various nutrients (which became more bioavailable).

Another group studied the effect of vermicompost of lantana on the grain yield and greenhouse gas (GHG) emissions from rice cultivation [53–55]. It was seen that fertilization by the weed's vermicompost led to better yields of rice, with significantly lesser emission of greenhouse gases than fertilization by chemicals [56,57].

So far, vermicomposting at a commercial scale has been largely confined to the use of animal manure as the feedstock. But animal manure has several competing uses, especially in developing countries such as India [10]. Consequently, it has limited supply as a vermicomposting feedstock. In contrast, weeds such as the ones explored in the present study have no competing use. They are more widely available, in much larger quantities, than animal manure. Secondly, the use of those weeds as vermireactor feedstock opens up the possibility of large-scale harvesting of such weeds. This, in turn, is likely to help in reducing the hold of those weeds in the areas dominated by them, enabling other vegetation to come up. Thirdly, the use of the weeds as vermireactor feedstock will prevent their debris and senescenced plants from degradation in the open, thereby preventing them from generating global warming gases. Lastly, organic fertilizers have high and unlimited demand. The use of weeds as feedstock can meet the demand. All these factors indicate the much higher economic viability of the present weed-based vermicomposting process than the pre-existing manure-based processes have.

4. Summary and Conclusions

A novel process has been reported which enables rapid, inexpensive, and sustainable vermicomposting of the toxic weeds parthenium *(Parthenium hysterophorus),* ipomoea *(Ipomoea carnea),* and lantana *(Lantana camara).* By invoking the concept of 'high-rate vermicomposting', developed earlier by S. A. Abbasi and coworkers, it has become possible to vermicompost the weeds directly without the need for pre-composting or providing any other form of pretreatment. The manure warm *Eisenia fetida*, which had been cultured on cowdung as feed, was slow to adapt to the weed-feed but survived and then began to thrive in all three weeds, enabling the sustained and efficient vermicomposting of the weeds throughout 480 days of uninterrupted operation of the vermireactors. In all cases, the extent of vermicast production per unit of time showed a rising trend, indicating that the rate of vermicomposting was set to rise further with time as the second and the third generations of earthworms, better adapted to the weeds than the pioneers, take over the feeding. The vermicomposting was found to accompany a $50 \pm 10\%$ loss of organic carbon of each weed. There was about an 8-fold increase in nitrate nitrogen reflecting the high degree of mineralization occurring in the course of vermicomposting. There was an equally dramatic increase in available sodium, potassium, calcium, and phosphorous. The levels of total copper, total manganese, and total zinc have also gone up mildly, while those of available copper, manganese, and zinc have gone up dramatically. There was a lowering of the carbon:nitrogen ratio to less than 15 in the vermicast of all three weeds, bringing the vermicast to the level considered highly desirable for use as fertilizer. The findings establish that sustained, direct, and rapid conversion of even toxic and allelopathic weeds to fertilizers can be accomplished with the high-rate vermicomposting paradigm. Among the three weeds, lantana was fed upon most voraciously by the earthworms, followed by parthenium and ipomoea. The juvenile and cocoon production was also the highest in lantana followed by ipomoea and parthenium.

Author Contributions: Conceptualization, S.A.A.; methodology, T.A., T.-A. and S.A.A.; software, T.-A.; validation, T.A. and T.-A.; formal analysis, S.A.A. and D.B.; investigation, T.-A., D.B. and T.A.; resources, S.A.A.; data curation, S.A.A. and T.-A.; writing—original draft preparation, T.-A. and T.A.; writing—review and editing, S.A.A.; visualization, D.B.; supervision, S.A.A.; project administration, S.A.A. and T.A.; funding acquisition, S.A.A. All authors have read and agreed to the published version of the manuscript.

Funding: The Council of Scientific and Industrial Research (CSIR), New Delhi, for the Emeritus Scientist grant (21(1034)/16/EMR-II).

Institutional Review Board Statement: Ethical review and approval were waived for this study because it involves no toxification, injuring, or harming any humans/animals in any manner.

Informed Consent Statement: Not applicable.

Data Availability Statement: Any data not reported in the paper will be provided on request.

Acknowledgments: SAA thanks the Council of Scientific and Industrial Research (CSIR), New Delhi, for the Emeritus Scientist grant (21(1034)/16/EMR-II).

Conflicts of Interest: The authors declare that there is no conflict of interest associated with the work described in this paper.

Abbreviations

HEVSTOW	High efficiency vertically stocked vermicomposting system for treating organic waste
PDCOP	Pseudo-discretized continuous reactor operation
SRT	Solid retention times
TOC	Total organic carbon
C:N	Carbon-to-nitrogen
SD	Standard deviation

References

1. Luque, G.M.; Bellard, C.; Bertelsmeier, C.; Bonnaud, E.; Genovesi, P.; Simberloff, D.; Courchamp, F. The 100th of the world's worst invasive alien species. *Biol. Invasions* **2014**, *16*, 981–985. [CrossRef]
2. Kumar, M.R.; Tauseef, S.M.; Abbasi, T.; Abbasi, S.A. Invasiveness and colonizing ability of ipomoea (*Ipomoea carnea*) and attempts at its management. *Nat. Environ. Pollut. Technol.* **2018**, *17*, 767–775.
3. Saha, P.; Mondal, A.; Sarkar, S. Phytoremediation of cyanide containing steel industrial wastewater by *Eichhornia crassipes*. *Int. J. Phytoremediat.* **2018**, *20*, 1205–1214. [CrossRef] [PubMed]
4. Hussain, N.; Abbasi, T.; Abbasi, S.A. Vermicomposting eliminates the toxicity of lantana (*Lantana camara*) and turns it into a plant friendly organic fertilizer. *J. Hazard. Mater.* **2015**, *298*, 46–57. [CrossRef]
5. Mungi, N.A.; Qureshi, Q.; Jhala, Y.V. Expanding niche and degrading forests: Key to the successful global invasion of *Lantana camara* (sensulato). *Glob. Ecol. Conserv.* **2020**, *23*, e01080. [CrossRef]
6. Kumar, S.; Varshney, J.G. Parthenium infestation and its estimated cost management in India. *Indian J. Weed Sci.* **2010**, *42*, 73–77.
7. Konishi, M.; Ito, M.; Ito, K. Weed invasion of forests caused by passage for recreational activities: A survey of Ashiu Forest. *Jpn. J. Conserv. Ecol.* **2009**, *14*, 45–54.
8. Shackleton, R.T.; Shackleton, C.M.; Kull, C.A. The role of invasive alien species in shaping local livelihoods and human well-being: A review. *J. Environ. Manag.* **2019**, *229*, 145–157. [CrossRef]
9. Shindell, D.T.; Faluvegi, G.; Koch, D.M.; Schmidt, G.A.; Unger, N.; Bauer, S.E. Improved attribution of climate forcing to emissions. *Science* **2009**, *326*, 716–718. [CrossRef]
10. Abbasi, S.A.; Nayeem-Shah, M.; Abbasi, T. Vermicomposting of phytomass: Limitations of the past approaches and the emerging directions. *J. Clean. Prod.* **2015**, *93*, 103–114. [CrossRef]
11. Tabassum-Abbasi; Patnaik, P.; Abbasi, S.A. Ability of Indian pennywort *Bacopa monnieri* (L.) Pennell in the phytoremediation of sewage (greywater). *Environ. Sci. Pollut. Res.* **2020**, *27*, 6078–6087. [CrossRef] [PubMed]
12. Tabassum-Abbasi; Patnaik, P.; Tauseef, S.M.; Abbasi, S.A. SHEFROL® bioreactor enhances the ability of pistia (*Pistia stratiotes*) in the phytoremediation of greywater. *Int. J. Environ. Anal. Chem.* **2020**, *102*, 1502–1511. [CrossRef]
13. Mohini, S.; Arpana, C.; Asokan, P. Flyash vermicompost from non-ecofriendly organic wastes. *Pollut. Res. Pap.* **1998**, *17*, 5–11.
14. Sannigrahi, A.K. Vermicomposting of some terrestrial weeds using *Perionyx Excavatus* earthworms. *Asian J. Microbiol. Biotechnol. Environ. Sci. Pap.* **2008**, *10*, 863–865.
15. Yadav, A.; Garg, V.K. Recycling of organic wastes by employing *Eisenia fetida*. *Bioresour. Technol.* **2011**, *102*, 2874–2880. [CrossRef]
16. Mahanta, K.; Jha, D.K.; Rajkhowa, D.J.; Manoj-Kumar. Microbial enrichment of vermicompost prepared from different plant biomasses and their effect on rice (*Oryza sativa* L.) growth and soil fertility. *Biol. Agric. Hortic.* **2012**, *28*, 241–250. [CrossRef]
17. Rajiv, P.; Rajeshwari, S.; Venckatesh, R. Bio-Fabrication of zinc oxide nanoparticles using leaf extract of Parthenium hysterophorus L. and its size-dependent antifungal activity against plant fungal pathogens. *Spectrochim. Acta Part A Mol. Biomol. Spectrosc.* **2013**, *112*, 384–387. [CrossRef]
18. Yadav, A.; Gupta, R.; Garg, V.K. Organic manure production from cow dung and biogas plant slurry by vermicomposting under field conditions. *Int. J. Recycl. Org. Waste Agric.* **2013**, *2*, 21. [CrossRef]
19. Suthar, S.; Sharma, P. Vermicomposting of toxic weed—*Lantana camara* biomass: Chemical and microbial properties changes and assessment of toxicity of end product using seed bioassay. *Ecotoxicol. Environ. Saf.* **2013**, *95*, 179–187. [CrossRef]
20. Rajiv, P.; Rajeshwari, S.; Rajendran, V. Impact of Parthenium weeds on earthworms (*Eudriluseugeniae*) during vermicomposting. *Environ. Sci. Pollut. Res.* **2014**, *21*, 12364–12371. [CrossRef]
21. Yadav, R.H. Assessment of different organic supplements for degradation of *Parthenium hysterophorus*by vermitechnology. *J. Environ. Health Sci. Eng.* **2015**, *13*, 44. [CrossRef] [PubMed]
22. Yadav, A.; Garg, V.K. Vermiconversion of biogas plant slurry and parthenium weed mixture to manure. *Int. J. Recycl. Org. Waste Agric.* **2016**, *5*, 301–309. [CrossRef]
23. Abbasi, T.; Tauseef, S.M.; Abbasi, S.A. The inclined parallel stack continuously operable vermireactor. *Off. J. Pat. Off.* **2011**, *22*, 9571.
24. Abbasi, S.A.; Banupriya, D.; Khamarang, C.; Abbasi, T. Initiatives and options for utilizing the dreaded weeds lantana (*Lantana camara*), parthenium (*Parthenium hysterophorus*) and salvinia (*Salvinia molesta*). *Int. J. Eng. Sci. Res.* **2018**, *6*, 95–145.
25. Tauseef, S.M.; Abbasi, T.; Banupriya, D.; Vaishnavi, V.; Abbasi, S.A. HEVSPAR: A novel vermireactor system for treating paper waste. *Off. J. Pat. Off.* **2013**, *24*, 12726.

26. Tauseef, S.M.; Abbasi, T.; Banupriya, G.; Banupriya, D.; Abbasi, S.A. A new machine for clean and rapid separation of vermicast, earthworms, and undigested substrate in vermicomposting systems. *Indian J. Environ. Health* **2014**, *56*, 485–488.
27. Hussain, N.; Tabassum-Abbasi Abbasi, T.; Abbasi, S.A. A Comparative study of the fertilizer-cum-pesticide effect of vermicomposts derived from cowdung and the toxic weed lantana (*Lantana camara*). *Agriculture* **2021**, *11*, 1263. [CrossRef]
28. Tauseef, S.M.; Tabassum-Abbasi; Patnaik, P.; Abbasi, T.; Abbasi, S.A. A novel high-rate vermicomposting machine for downstream processing of biorefinery waste and other forms of organic solid waste. *Biomass Convers. Biorefin.* **2021**. [CrossRef]
29. Nayeem-Shah, M.; Gajalakshmi, S.; Abbasi, S.A. Direct, rapid and sustainable vermicomposting of the leaf litter of neem (*Azadirachta indica*). *Appl. Biochem. Biotechnol.* **2015**, *2*, 792–801. [CrossRef]
30. Carter, M.R.; Gregorich, E.G. (Eds.) *Soil Sampling and Methods of Analysis*, 2nd ed.; CRC Press, Taylor & Francis Group: Boca Raton, FL, USA, 2008; p. 1224.
31. Bashour, I.I.; Sayegh, A.H. *Methods of Analysis for Soils of Arid and Semi-Arid Regions*; FAO: Rome, Italy, 2007.
32. Heanes, D.L. Determination of total organic-C in soils by an improved chromic acid digestion and spectrophotometric procedure. *Commun. Soil Sci. Plant Anal.* **1984**, *15*, 1191–1213. [CrossRef]
33. Kandeler, E. Bestimung von Gesamtstickstoffnachkjeldahl. In *Bodenbiologische Arbeitsmethoden*; Schinner, F., Kandeler, E., Ohlinger, R., Margesin, R., Eds.; Springer: Berlin/Heidelberg, Germany, 1993; pp. 346–366.
34. Jones, J.B. *Laboratory Guide for Conducting Soil Tests and Plant Analysis*; CRC Press LLC: New York, NY, USA, 2001.
35. Mehlich, A. Mehlich 3 soil test extractant: A modification of Mehlich 2 extractant. *Commun. Soil Sci. Plant Anal.* **1984**, *15*, 1409–1416. [CrossRef]
36. Knudsen, D.; Beegle, D. Recommended Chemical Soil Test Procedures for the North Central Region: # 6. In *Recommended Phosphorus Tests*; North Central Regional Publication: Kansas City, KS, USA, 1998; Volume 221.
37. Edwards, C.A.; Subler, S.; Arancon, N.Q. *Vermiculture Technology—Earthworms, Organic Wastes and Environmental Management*; CRC Press: Boca Raton, FL, USA, 2011.
38. Ndegwa, P.M.; Thompson, S.A.; Das, K.C. Effects of stocking density and feeding rate on vermicomposting of biosolids. *Bioresour. Technol.* **2000**, *71*, 5–12. [CrossRef]
39. Deka, H.; Deka, S.; Baruah, C.K.; Das, J.; Hoque, S.; Sarma, H.; Sarma, N.S. Vermicomposting potentiality of Perionyx excavatus for recycling of waste biomass of java citronella-An aromatic oil yielding plant. *Bioresour. Technol.* **2011**, *102*, 11212–11217. [CrossRef] [PubMed]
40. Hussain, N.; Abbasi, T.; Abbasi, S.A. Vermiremediation of an invasive and pernicious weed salvinia (*Salvinia molesta*). *Ecol. Eng.* **2016**, *91*, 432–440. [CrossRef]
41. Hussain, N.; Abbasi, T.; Abbasi, S.A. Transformation of toxic and allelopathic lantana into a benign organic fertilizer through vermicomposting. *Spectrochim. Acta Part A Mol. Biomol. Spectrosc.* **2016**, *163*, 162–169. [CrossRef]
42. Hussain, N.; Abbasi, T.; Abbasi, S.A. Vermicomposting-mediated conversion of the toxic and allelopathic weed ipomoea into a potent fertilizer. *Process Saf. Environ. Prot.* **2016**, *103*, 97–106. [CrossRef]
43. Patnaik, P.; Abbasi, T.; Abbasi, S.A. Vermicompost of the widespread and toxic xerophyte prosopis (*Prosopis juliflora*) is a benign organic fertilizer. *J. Hazard. Mater.* **2020**, *399*, 122864. [CrossRef]
44. Hussain, N.; Abbasi, T.; Abbasi, S.A. Vermicomposting transforms allelopathic parthenium into a benign organic fertilizer. *J. Environ. Manag.* **2016**, *180*, 180–189. [CrossRef]
45. Hussain, N.; Abbasi, T.; Abbasi, S.A. Transformation of the pernicious and toxic weed parthenium into an organic fertilizer by vermicomposting. *Int. J. Environ. Stud.* **2016**, *73*, 731–745. [CrossRef]
46. Hussain, N.; Abbasi, T.; Abbasi, S.A. Generation of highly potent organic fertilizer from pernicious aquatic weed *Salvinia molesta*. *Environ. Sci. Pollut. Res.* **2018**, *25*, 4989–5002. [CrossRef]
47. Hussain, N.; Abbasi, T.; Abbasi, S.A. Evaluating the fertilizer and pesticidal value of vermicompost generated from a toxic and allelopathic weed ipomoea. *J. Saudi Soc. Agric. Sci.* **2020**, *19*, 43–50. [CrossRef]
48. Hussain, N.; Abbasi, S.A. Efficacy of the vermicomposts of different organic wastes as "clean" fertilizers: State-of-the-art. *Sustainability* **2018**, *10*, 1205. [CrossRef]
49. Hussain, N.; Abbasi, T.; Abbasi, S.A. Toxic and allelopathic ipomoea yields plant-friendly organic fertilizer. *J. Clean. Prod.* **2017**, *148*, 826–835. [CrossRef]
50. Hussain, N.; Abbasi, T.; Abbasi, S.A. Enhancement in the productivity of ladies finger (*Abelmoschus esculentus*) with concomitant pest control by the vermicompost of the weed salvinia (*Salvinia molesta*, Mitchell). *Int. J. Recycl. Org. Waste Agric.* **2017**, *6*, 335–343. [CrossRef]
51. Hussain, N.; Abbasi, T.; Abbasi, S.A. Detoxification of parthenium (*Parthenium hysterophorus*) and its metamorphosis into an organic fertilizer and biopesticide. *Bioresour. Bioprocess.* **2017**, *4*, 26. [CrossRef]
52. Patnaik, P.; Abbasi, T.; Abbasi, S.A. Fate of chemical constituents of prosopis leaves during the conversion of the leaves into an organic fertiliser. *Int. J. Environ. Anal. Chem.* **2021**. [CrossRef]
53. Dhanuja, C. Greenhouse Gas Emissions from Paddy Fields with Special Reference to the Effect of Vermicomposts and Synthetic Chemical Fertilizers. Ph.D. Thesis, Pondicherry University, Puducherry, India, 2019; pp. xiii+249.
54. Dhanuja, C.; Saxena, D.K.; Abbasi, T.; Abbasi, S.A. Effect of application of vermicompost on methane emission and grain yield of *Chinna Ponni* paddy crop. *Paddy Water Environ.* **2019**, *17*, 797–802. [CrossRef]

55. Dhanuja, C.; Abbasi, T.; Abbasi, S.A. Fertilization of paddy cultivation with vermicompost: A critical mini review. *Org. Agric.* **2020**, *10*, 309–325. [CrossRef]
56. Tabassum-Abbasi; Abbasi, T.; Luithui, C.; Abbasi, S.A. Modelling methane and nitrous oxide emissions from rice paddy wetlands in India using Artificial Neural Networks (ANNs). *Water* **2019**, *12*, 751.
57. Tabassum-Abbasi; Abbasi, T.; Luithui, C.; Abbasi, S.A. A Model to Forecast Methane Emissions from Tropical and Subtropical Reservoirs on the Basis of Artificial Neural Networks. *Water* **2020**, *12*, 145. [CrossRef]

agriculture

Article

Vermicompost Amendment in Soil Affects Growth and Physiology of *Zea mays* Plants and Decreases Pb Accumulation in Tissues

Zaiga Landorfa-Svalbe, Māra Vikmane and Gederts Ievinsh *

Department of Plant Physiology, Faculty of Biology, University of Latvia, 1 Jelgavas Str., LV-1004 Riga, Latvia
* Correspondence: gederts.ievins@lu.lv

Abstract: Minimization of the possible harmful effects of soil pollution on agricultural production and food safety are the major challenges in modern agriculture. There is great scientific interest in the detailed understanding of the physiology of lead uptake and toxicity in *Zea mays*, together with the search for approaches to minimizing Pb accumulation in tissues. The aim of the present study was to explore the possibility of reducing Pb accumulation in *Z. mays* plants cultivated in Pb-contaminated soil, by means of vermicompost amendment. *Z. mays* plants were cultivated at three soil vermicompost amendment rates (10, 20, and 30%), with the addition of 1000 mg L^{-1} of Pb in the form of $Pb(NO_3)_2$ or an equivalent amount of nitrogen in the form of NH_4NO_3. Additional nitrogen had a significant stimulatory effect on plant growth and physiology, but only for control plants, and at a low vermicompost amendment rate. Independently, Pb had an insignificant negative effect on plant growth and biomass partitioning, but significantly negatively affected the mineral nutrition of *Z. mays* plants. At a 10 and 20% soil vermicompost amendment rate, the Pb concentration in plant leaves and roots decreased by 65%, while plant biomass increased four to five times in comparison to soil-grown control plants, together with accelerated flowering. It was concluded that vermicompost is one of the most promising soil amendments for reducing heavy metal uptake and accumulation in crop plants, while also being an efficient organic fertilizer.

Keywords: chlorophyll; lead; maize; nitrate; nitrogen; organic fertilizer; physiological status

1. Introduction

Agricultural practices leading to increased soil sustainability and food safety are gaining more interest from both scientists and farmers [1,2]. Minimization of the possible harmful effects of soil pollution on agricultural production and food safety are the major challenges in this respect. While the problem of the negative effect of soil heavy metal contamination on yields can be solved by developing metal-tolerant crop varieties [3], the presence of toxic metal levels in agricultural crop products still represents an insufficiently addressed problem [4].

Lead is one of the environmental contaminants that is mostly associated with anthropogenic impacts [5]. Lead bioavailability in soils is controlled by complex interactions, as Pb is readily complexed with both inorganic constituents as well as organic ligands, or adsorbed on the surface of different types of particles [6]. Therefore, different soil amendments have been explored for their ability to adsorb Pb and other heavy metals and, thus, decrease their bioavailability. Vermicompost is an especially promising product for soil stabilization of heavy metals, as it also acts as a valuable organic fertilizer. The primary positive effect of vermicompost on plant growth is related to the high content of plant-available mineral nutrients as well as plant-growth-promoting biologically active substances—as reported in a recent review [7].

The majority of adsorption studies with vermicompost and heavy metals have been performed in laboratory conditions without plants [8–11]. Adsorption studies with vermicompost using plants in controlled conditions are relatively scarce [12,13]. However, there have been some field experiments on the effect of organic amendments on heavy metal accumulation [14,15].

Zea mays L. is a crop species with a relatively high potential for use in phytoremediation [16]. There is still a great scientific interest in the detailed understanding of the physiology of lead uptake and toxicity in *Z. mays*, as evidenced by a recent review [17]. It appears that studies aiming at understanding the amendment methods and functional mechanisms leading to reduced Pb uptake and accumulation are critically relevant.

Direct deleterious effects of heavy metals in plants, including Pb, often have not been distinguished from controlled physiological responses. Even non-biogenous heavy metals can induce changes in gene expression patterns due to chemical similarity with essential metals [18]. The downregulation of photosynthesis and resource allocation to defense due to heavy metal treatment are typical cases of induced responses associated with physiological alterations [19]. Alternatively, endogenous oxidative stress through the enhanced formation of reactive oxygen species can lead to the inhibition of photosynthesis in heavy-metal-stressed plants [20]. Therefore, non-destructive methods of physiological measurement, such as chlorophyll analysis and chlorophyll *a* fluorescence assays, can be used as indicators of the physiological status of plants [21], including in studies associated with mineral nutrient availability [22,23].

One of problems in studies aiming at assessing the effects of Pb on plants is related to the fact that lead nitrate is very often used for treatments, without an additional control in the form of a balancing nitrogen fertilizer [13]. As a result, any growth stimulation or other physiological changes by lead nitrate can be erroneously interpreted as the effect of Pb itself [24,25].

The aim of the present study was to explore the possibility of reducing Pb accumulation in *Z. mays* plants cultivated in Pb-contaminated soil, by means of vermicompost amendment. In addition, the effect of vermicompost for growth improvement during *Z. mays* cultivation was assessed. Special care was taken to evaluate the possible effect of increased soil nitrogen content due to treatment with high doses of Pb nitrate.

2. Materials and Methods

2.1. Plant Material and Substrates

Seeds of *Zea mays* L. var. *saccharata* cv. 'Zlota Karlova SNF' (Toraf, Kujakowice Górne, Poland) were used for the experiment. The variety is a very early, dwarf-type sugar corn.

Agricultural soil (loamy sand, 2% organic matter) collected in October from a field (Valmiera Municipality, Latvia) employed for cereal production was used as a substrate for plant cultivation. Analysis of plant-available mineral nutrient concentration in soil was performed in a certified agrochemical laboratory (Table 1). According to the established criteria [26], the soil was relatively rich in plant-available nutrients. The soil Pb concentration was <2.3 mg kg^{-1}.

Vermicompost (Eko Zeme, Bauska District, Latvia) was purchased from a local supplier. The vermicompost was produced from composted cow manure and grass biomass, and was certified for organic agriculture. The analysis of plant-available mineral nutrient concentration in vermicompost was performed in a certified agrochemical laboratory (Table 1). The used vermicompost was a very good source of plant-available N, P, K, Mg, Mn, and Zn. The vermicompost Pb concentration was <2.3 mg kg^{-1}.

Table 1. Properties of agricultural soil and vermicompost used in the present study.

Nutrient or Property (Unit)	Soil	Vermicompost	Optimum for Cultivated Plants
N (mg L^{-1})	90	730	120
P (mg L^{-1})	316	4251	60
K (mg L^{-1})	560	16,500	150
Ca (mg L^{-1})	1700	25,000	800
Mg (mg L^{-1})	320	4500	50
S (mg L^{-1})	23	925	50
Fe (mg L^{-1})	925	420	30
Mn (mg L^{-1})	145	165	1.5
Zn (mg L^{-1})	11	80	1.0
Cu (mg L^{-1})	2.75	6.00	0.50
Mo (mg L^{-1})	0.09	0.04	0.02
B (mg L^{-1})	1.1	3.0	0.2
Na (mg L^{-1})	32	780	n.a.
pH_{KCl} (pH units)	5.87	7.29	n.a.
Electrical Conductivity (mS m^{-1})	1.84	36.7	n.a.

Plant-available concentrations are indicated, measured in 1 M HCl extract. Electrical conductivity was measured at 1:5 extraction ratio.

2.2. Plant Cultivation and Treatments

The experiments were performed in winter in an experimental greenhouse with an automatic control system (HortiMaX, Maasdijk, The Netherlands). Additional light was supplemented by Master SON-TPIA Green Power CG T 400 W (Philips, Amsterdam, The Netherlands) and Powerstar HQI-BT 400 W/D PRO (Osram, Munich, Germany) lamps (380 µmol m^{-2} s^{-1} at the plant level) for a 16 h photoperiod, with a day/night temperature 25/16 °C and relative air humidity of 60 to 70%.

The substrate for cultivation was prepared from air-dried samples of soil and vermicompost at three vermicompost amendment rates: 10, 20, and 30% (*v*/*v*). The control treatment contained only soil. Deionized water was added to achieve 50% substrate moisture, measured with a HH2 moisture meter equipped with a WET-2 sensor (Delta-T Devices, Burwell, Cambridge, UK). For each of four substrate vermicompost amendment rates, there were three subtreatments: control, $Pb(NO_3)_2$, and NH_4NO_3 (Table 2). The amount of $Pb(NO_3)_2$ applied per 1 L of substrate (1598 mg) gave a content of 1000 mg L^{-1} of Pb. Accordingly, 388 mg of NH_4NO_3 per 1 L of substrate had the amount of N equivalent to that in the $Pb(NO_3)_2$ treatment. Necessary amounts of both salts were dissolved in deionized water and applied to the respective substrate in the form of a 10% solution. Prepared substrates were placed in 1.2 L plastic containers, 1 L per container.

Seeds were surface-disinfected with 1% $KMnO_4$ solution, rinsed 10 times with deionized water, and imbibed for 6 h in water. Seeds were placed on filter paper in Petri dishes and germinated in the dark for 5 days at 20 °C. Well-developed germinated seeds were individually sown in containers with the prepared substrate at a 1 cm depth, 10 seeds per treatment. Containers were randomly placed in the greenhouse. After one week, typical uniform seedlings, five per treatment, were selected for further cultivation. Plants were watered with deionized water to maintain substrate moisture at the 50–60% level. Once a week, individual containers were randomly repositioned on a greenhouse bench.

Table 2. Treatments used in the present study.

Code	Vermicompost (%, *v/v*)	$Pb(NO_3)_2$ (mg L^{-1})	NH_4NO_3 (mg L^{-1})
V0	0	0	0
V0 + N	0	0	388
V0 + Pb	0	1598	0
V10	10	0	0
V10 + N	10	0	388
V10 + Pb	10	1598	0
V20	20	0	0
V20 + N	20	0	388
V20 + Pb	20	1598	0
V30	30	0	0
V30 + N	30	0	388
V30 + Pb	30	1598	0

2.3. Measurements and Termination

To monitor plant growth, starting from week 3, the height of individual plants was measured weekly, from the base of the stem to the tip of the longest leaf.

Starting from week 4, nondestructive physiological measurements were performed weekly. For each individual plant, three of the upper photosynthetically most active leaves were selected for measurements. Leaf chlorophyll concentration was measured using a chlorophyll meter CCM-300 (Opti-Sciences, Hudson, NH, USA). Chlorophyll *a* fluorescence was measured in leaves dark-adapted for at least 20 min by a Handy PEA fluorometer (Hansatech Instruments, Pentney, King's Lynn, UK). Fluorescence data analysis was performed by PEA Plus software (Hansatech Instruments, Pentney, King's Lynn, UK). The photochemical efficiency of photosynthesis was estimated by the multiparametric fluorescence indicator, Performance Index Total, combining information on the status of both photosystems, as well as the electron flow between the two systems on an absorption basis [27].

The experiment was terminated after nine weeks of cultivation, when plants in all treatments started to develop male inflorescences. Stem height up to the inflorescence base was measured. Individual plants were cut at the substrate level and separated into individual parts: dry leaves, live leaves, inflorescence, and stem. Roots were separated from the substrate and washed under running tap water to remove any adhered particles, rinsed with deionized water, and blotted dry with paper towels. The fresh mass of individual parts was measured and tissues were dried in a oven at 60 °C until no change in biomass occurred; then, dry mass was measured. Water content in the plant tissues was calculated as g H_2O per g dry mass.

2.4. Analytical Measurements

Dried plant material was used for measurement of soluble K^+ and NO_3^- concentration in the flag leaf, base leaf, stem base, and roots. Plant material (about 5 g) was crushed in pieces and homogenized, and a sample of 0.2 g was randomly taken from the plant material. Tissues were ground with a mortar and pestle to a fine powder and 10 mL of deionized water was added. After filtration through a nylon mesh cloth (No. 80), homogenate was used for measurement of the K^+ concentration by a LAQUAtwin compact meter B-731 and NO_3^- concentration by a LAQUAtwin compact meter NO3-11 (Horiba, Kyoto, Japan). For nitrate measurement, a nitrate interference suppressor solution (Mettler-Toledo, Schwerzenbach, Switzerland) was used according to the manufacturer's instructions. For each treatment, three samples from individual plants were measured in at least three analytical replicates and the average value was calculated.

Pb analysis was performed in a certified analytical laboratory. Briefly, H_2O_2 and HNO_3 were added to 0.3 g of homogenized plant material. The samples were digested using a microwave digestion system Mars 6 (CEM Corporation, Matthews, NC, USA). Pb was measured using an Agilent 7700 Series ICP-MS (Agilent Technologies, Santa Clara, CA, USA). Five replicate samples for leaves and three replicate samples for roots were analyzed for each treatment.

2.5. Data Analysis

The results were analyzed by KaleidaGraph (v. 5.0, Synergy Software, Reading, PA, USA). The statistical significance of differences was evaluated by one-way ANOVA using post hoc analysis (Tukey's HSD).

3. Results

3.1. Effect on Growth

Changes in the growth of *Z. mays* plants due to vermicompost amendment and nitrogen and Pb treatment were estimated by weekly measurements of plant height. The treatment of control plants with NH_4NO_3 resulted only in a temporary increase in plant height (Figure 1A). The greatest positive effect of vermicompost amendment on *Z. mays* growth was observed for the 10% vermicompost (Figure 1B) and 20% vermicompost treatments (Figure 1C). However, the growth of plants amended with 30% vermicompost was initially suppressed in comparison to control plants (Figure 1D). The nitrogen and Pb nitrate treatments did not result in a growth response in comparison to the control. However, the final height of plants clearly showed that the nitrogen treatment of *Z. mays* plants amended with 30% vermicompost resulted in a statistically significant negative effect (Figure 2). A similar effect was observed for changes in shoot dry biomass (Figure 3).

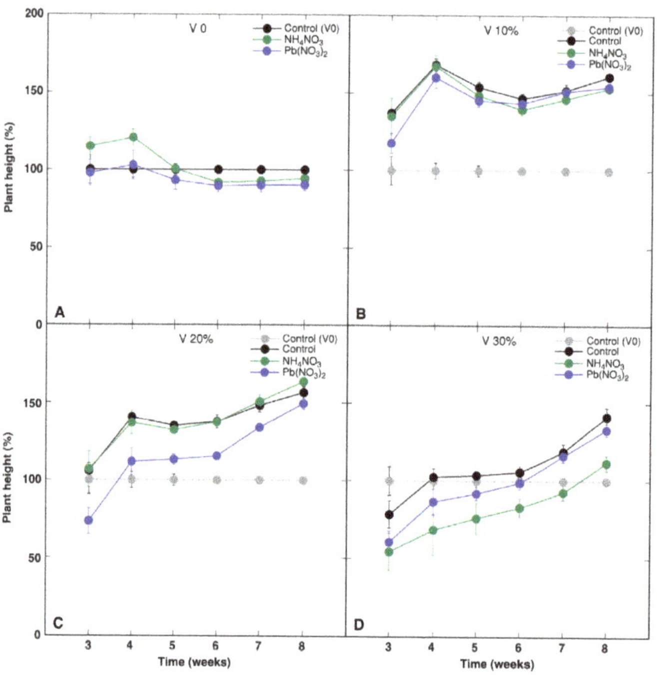

Figure 1. Effect of rate of soil vermicompost (V) amendment, NH_4NO_3, and $Pb(NO_3)_2$ treatment on relative growth of *Zea mays* plants grown in soil (**A**), and at soil vermicompost amendment rates of 10% (**B**), 20% (**C**), and 30% (**D**). Amendment rate is given as *v/v* % added to soil. Data are means from 5 replicates ± SE. V 0: no vermicompost amendment; V 10%, V 20%, and V 30%: vermicompost amendment by 10, 20, and 30% (*v*/*v*), respectively.

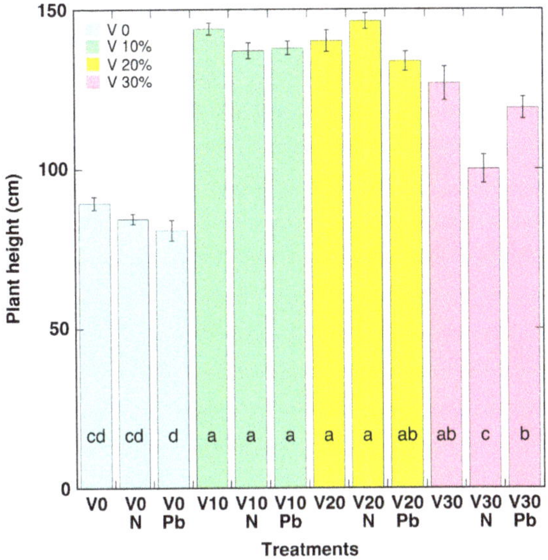

Figure 2. Effect of rate of soil vermicompost (V) amendment, NH_4NO_3 (N), and $Pb(NO_3)_2$ (Pb) treatment on final height of *Zea mays* plants. Amendment rate is given as *v*/*v*% added to soil. Data are means from 5 replicates ± SE. Different letters indicate statistically significant differences between the treatments according to Tukey's HSD test ($p < 0.05$).

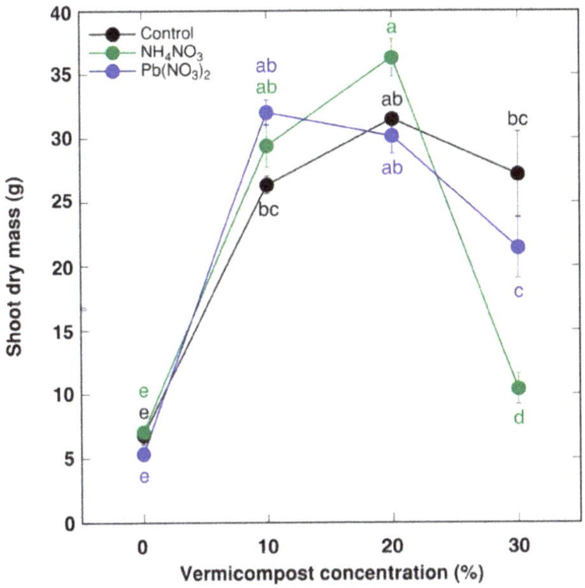

Figure 3. Effect of rate of soil vermicompost amendment, NH_4NO_3, and $Pb(NO_3)_2$ treatment on shoot dry mass of *Zea mays* plants. Amendment rate is given as *v*/*v* % added to soil. Data are means from 5 replicates ± SE. Different letters indicate statistically significant differences between the treatments according to Tukey's HSD test ($p < 0.05$).

Biomass partitioning in *Z. mays* plants indicated that the growth of all plant parts was stimulated by vermicompost amendment, but to a different extent (Figure 4). The highest degree of stimulation at 10 and 20% vermicompost amendment rates occurred for flowers (Figure 4A) followed by stems (Figure 4B), but also the biomass of leaves (Figure 4C) and roots (Figure 4D) significantly increased. At the 10% vermicompost amendment rate, the biomass of all parts of the plants treated with nitrogen and Pb nitrate tended to be higher in comparison to the control plants, but the differences were not statistically significant. In general, plants amended with 30% vermicompost had a lower biomass of their parts, but significant growth reduction was evident for the leaves, stems, and roots of *Z. mays* plants treated with nitrogen and the roots of plants treated with Pb nitrate.

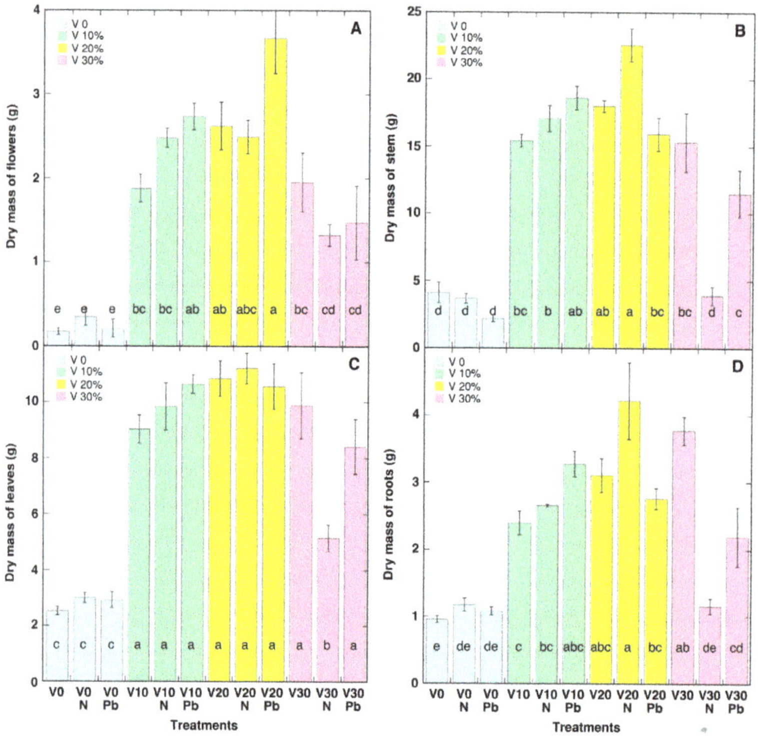

Figure 4. Effect of rate of soil vermicompost (V) amendment, NH_4NO_3 (N), and $Pb(NO_3)_2$ (Pb) treatment on dry mass of leaves (**A**), dry mass of stem (**B**), dry mass of flowers (**C**), and dry mass of roots (**D**) of *Zea mays* plants. Amendment rate is given as *v/v* % added to soil. Data are means from 5 replicates ± SE. Different letters indicate statistically significant differences between the treatments according to Tukey's HSD test ($p < 0.05$).

The total number of leaves was not significantly affected by vermicompost amendment and nitrogen and Pb nitrate treatments (data not shown). However, the number of dry leaves significantly decreased (Figure 5A) and that of live leaves significantly increased (Figure 5B) for *Z. mays* plants growing in soil amended with 30% vermicompost.

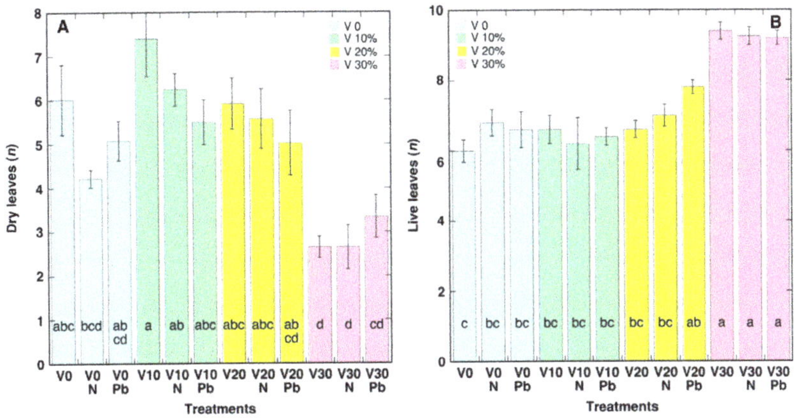

Figure 5. Effect of rate of soil vermicompost (V) amendment, NH_4NO_3 (N), and $Pb(NO_3)_2$ (Pb) treatment on number of dry leaves (**A**) and number of live leaves (**B**) of *Zea mays* plants. Amendment rate is given as *v/v* % added to soil. Data are means from 5 replicates ± SE. Different letters indicate statistically significant differences between the treatments according to Tukey's HSD test ($p < 0.05$).

The water content increased in photosynthetically active leaves of *Z. mays* control plants treated with nitrogen and Pb nitrate, in Pb nitrate-treated plants that received 20% vermicompost amendment, and for all plants in the 30% vermicompost treatment (Figure 6A). Moreover, the water content in stems significantly increased in Pb-nitrate-treated plants for the 20% vermicompost amendment, and in all plants for the 30% vermicompost amendment, and this effect was especially pronounced for nitrogen-treated plants (Figure 6B).

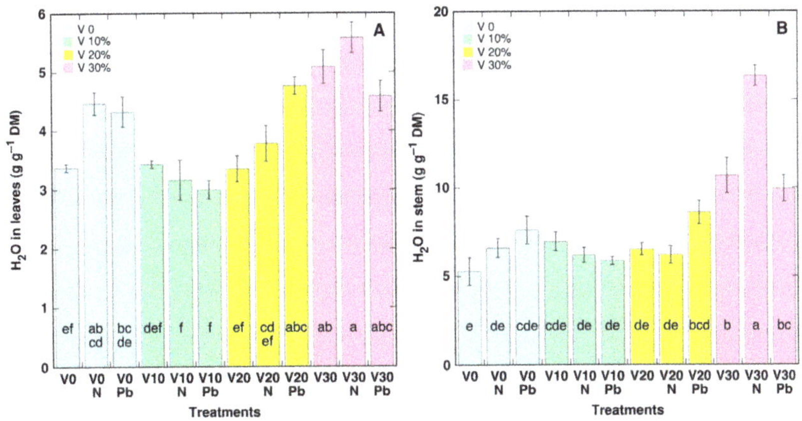

Figure 6. Effect of rate of soil vermicompost (V) amendment, NH_4NO_3 (N), and $Pb(NO_3)_2$ (Pb) treatment on water content in leaves (**A**) and water content in stems (**B**) of *Zea mays* plants. Amendment rate is given as *v/v* % added to soil. Data are means from 5 replicates ± SE. Different letters indicate statistically significant differences between the treatments according to Tukey's HSD test ($p < 0.05$).

3.2. Effect on Physiological Parameters

For soil-grown plants *Z. mays*, chlorophyll concentration in the main photosynthesizing leaves of the plants treated with nitrogen and Pb nitrate started to increase over control values on Week 6 (Figure 7A). Soil amendment with vermicompost at a 10% rate resulted in a significant increase in leaf chlorophyll concentration, with an additional increase due to treatment with nitrogen and Pb nitrate (Figure 7B). However, at higher vermicompost amendment rates, differences between control plants and those treated with nitrogen and Pb nitrate levelled off (Figure 7C) and completely disappeared (Figure 7D).

There was a temporary increase in chlorophyll *a* fluorescence parameter Performance Index Total in soil-grown plants in the nitrogen and Pb nitrate treatment (Figure 8A). The increase in Performance Index Total became more pronounced and continuous with increased vermicompost amendment rate (Figure 8B–D). Treatment of *Z. mays* plants with nitrogen and Pb nitrate at the 10 and 20% vermicompost substitution rate resulted in an additional increase in Performance Index Total, but this effect was no longer evident at the 30% vermicompost amendment rate.

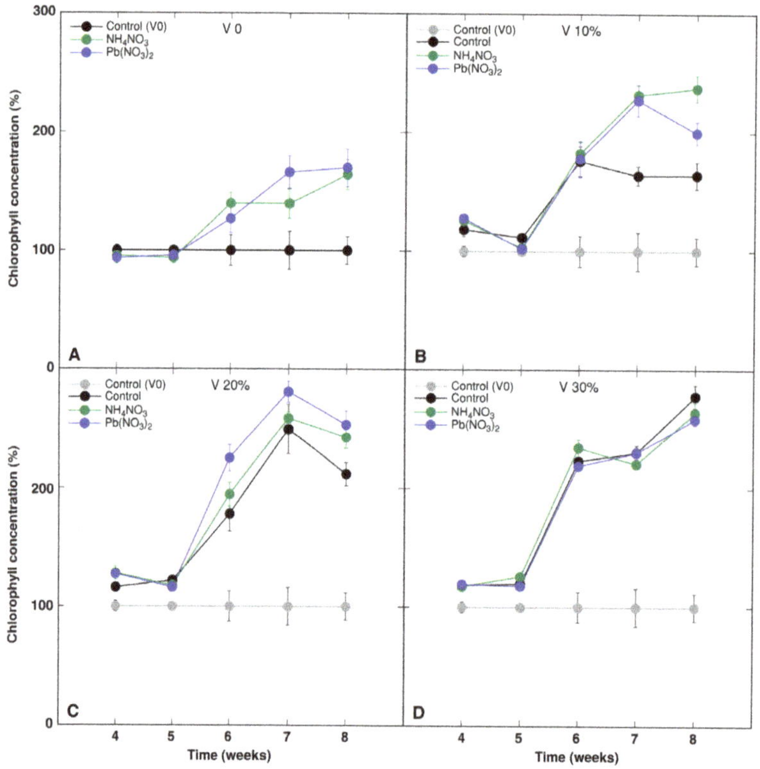

Figure 7. Effect of NH_4NO_3 (N) and $Pb(NO_3)_2$ (Pb) treatment on relative time course of leaf chlorophyll concentration of *Zea mays* plants grown in soil (**A**), and at soil vermicompost amendment rates of 10% (**B**), 20% (**C**), and 30% (**D**). Amendment rate is given as *v/v* % added to soil. Data are means from 5 replicates ± SE, with 3 separate measurements each. V 0: no vermicompost amendment; V 10%, V 20%, and V 30%: vermicompost amendment by 10, 20, and 30% (*v*/*v*), respectively.

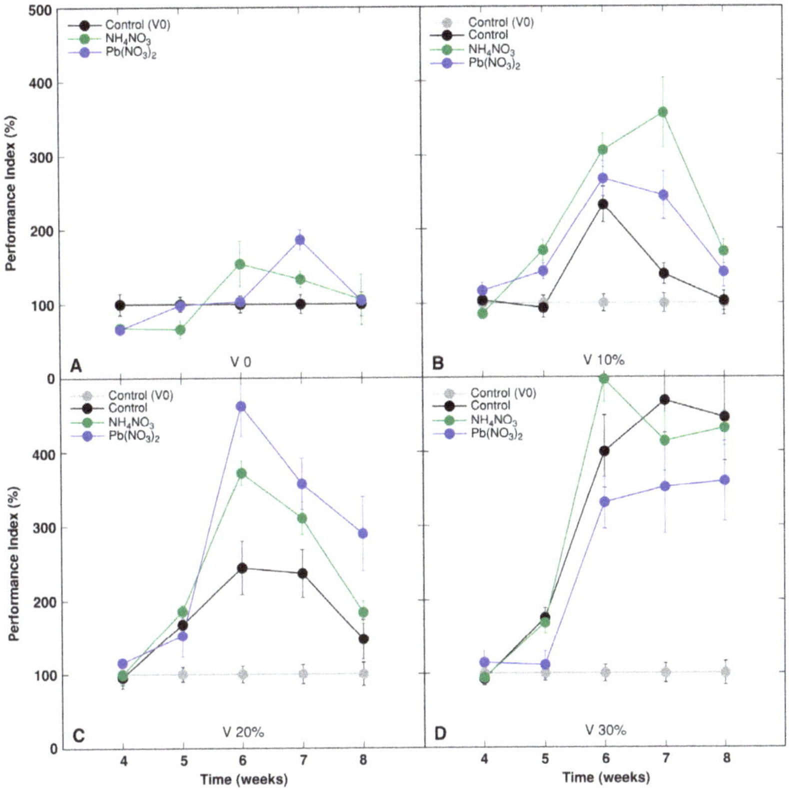

Figure 8. Effect of NH_4NO_3 (N) and $Pb(NO_3)_2$ (Pb) treatment on relative time course of chlorophyll *a* fluorescence parameter Performance Index Total of *Zea mays* plants grown in soil (**A**), and at soil vermicompost amendment rates of 10% (**B**), 20% (**C**), and 30% (**D**). Amendment rate is given as *v/v* % added to soil. Data are means from 5 replicates ± SE, with 3 separate measurements each. V 0: no vermicompost amendment; V 10%, V 20%, V 30%: vermicompost amendment by 10, 20, and 30% (*v*/*v*), respectively.

3.3. Effect on Accumulation of Ions and Pb

Nitrate concentration in different plant parts was measured as a possible indicator of nitrogen status. Nitrate concentration in flag leaf tissue was not affected by the treatments (Figure 9A). Surplus nitrate accumulated in the base leaf of plants cultivated at the 30% vermicompost amendment rate, and especially for plants treated with additional nitrogen and Pb nitrate (Figure 9B). Similarly, nitrate accumulated in the stem base of *Z. mays* plants treated with additional nitrate, and this effect increased with increasing soil vermicompost amendment rate, but Pb drastically reduced the nitrogen-dependent increase in nitrate accumulation (Figure 9C). Root nitrate concentration significantly increased due to the increase in soil vermicompost amendment rate, and it was stimulated by additional treatment with nitrogen at the 30% amendment rate (Figure 9D). However, Pb treatment resulted in a significant reduction in nitrate accumulation in roots.

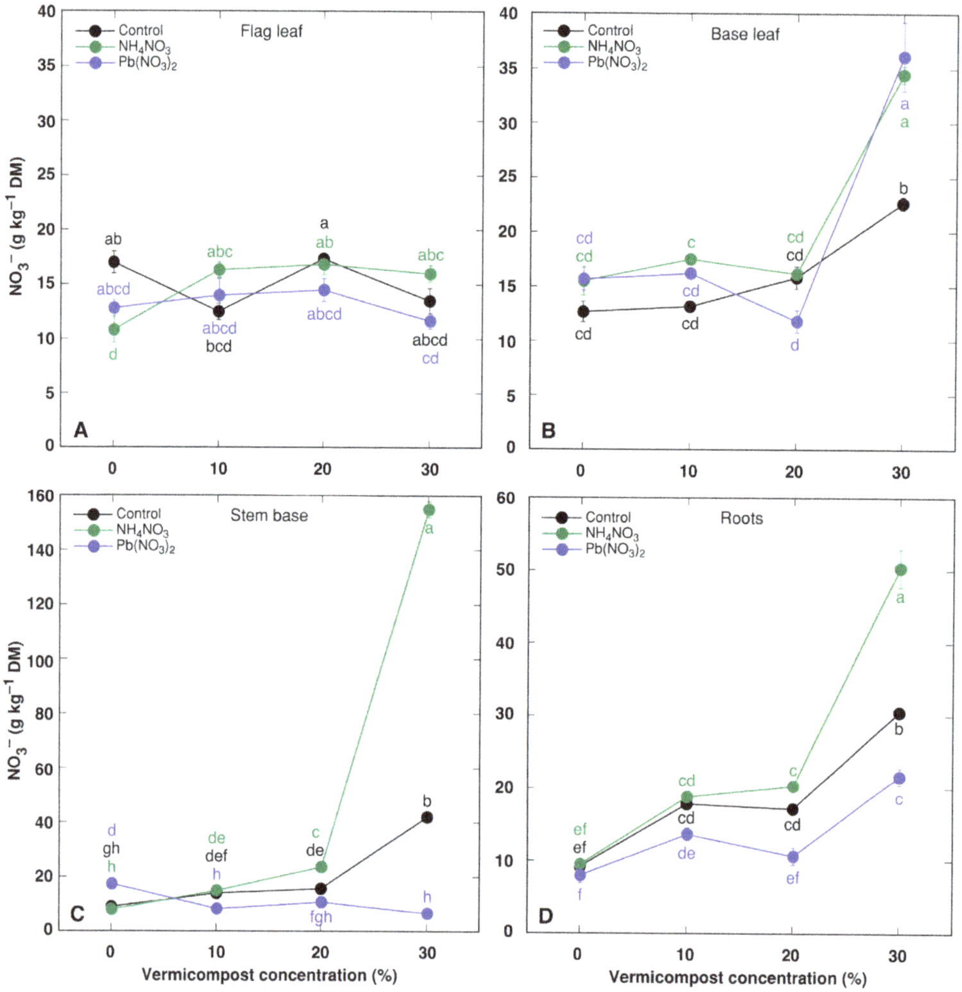

Figure 9. Effect of rate of soil vermicompost amendment, NH_4NO_3, and $Pb(NO_3)_2$ treatment on NO_3^- concentration in flag leaf (**A**), base leaf (**B**), stem base (**C**), and roots (**D**) of *Zea mays* plants. Amendment rate is given as *v/v* % added to soil. Data are means from 3 replicates ± SE. Different letters indicate statistically significant differences between the treatments according to Tukey's HSD test ($p < 0.05$). DM, dry mass.

The effect of vermicompost amendment and addition treatments on general plant nutrition was evaluated by tissue K^+ concentration in different plant parts (Figure 10). The effect of additional plant-available K^+ through soil amendment with vermicompost was clearly observed for all tested plant parts, and particularly in the stem base and base leaf. A stimulative effect of additional nitrogen availability on K^+ accumulation was the most pronounced in flag leaf tissues, together with a striking negative effect of Pb (Figure 10A), which was evident also in stem base tissue (Figure 10C).

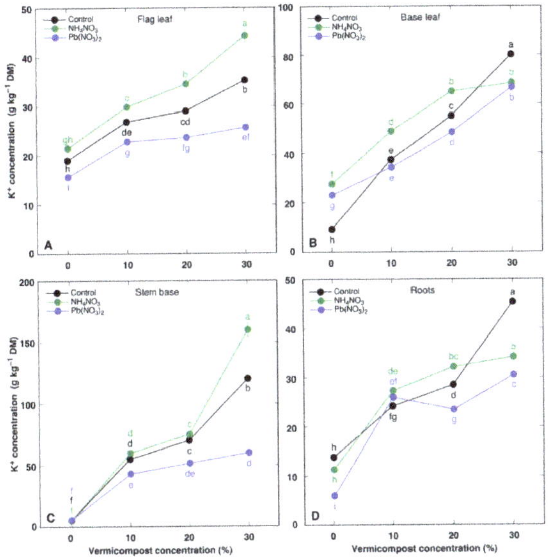

Figure 10. Effect of rate of soil vermicompost amendment, NH_4NO_3, and $Pb(NO_3)_2$ treatment on K^+ concentration in flag leaf (**A**), base leaf (**B**), stem base (**C**), and roots (**D**) of *Zea mays* plants. Amendment rate is given as *v/v* % added to soil. Data are means from 3 replicates ± SE. Different letters indicate statistically significant differences between the treatments according to Tukey's HSD test ($p < 0.05$). DM, dry mass.

The concentration of Pb was 1.34 and 3.01 mg kg^{-1} in the leaves and roots of plants grown in uncontaminated soil, respectively. In plants treated with 1 g L^{-1} of Pb in the form of nitrate, these levels reached 100 and 500 mg kg^{-1}, respectively (Figure 11). However, soil amendment with vermicompost significantly decreased Pb accumulation approximately by 80%. This decrease in Pb accumulation with increased vermicompost amendment rate occurred in both leaves and roots (Figure 11, inset).

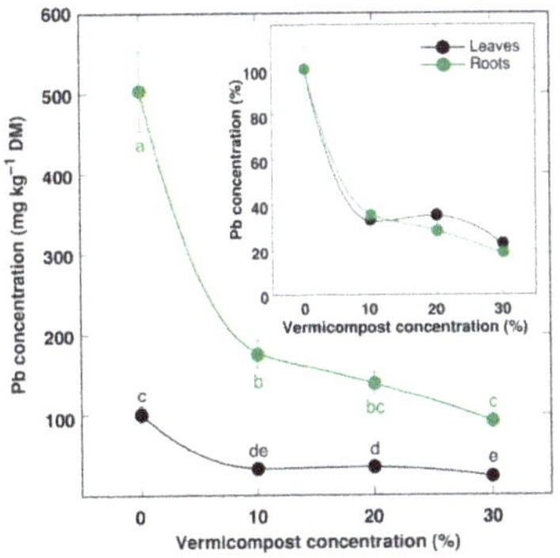

Figure 11. Effect of rate of soil vermicompost amendment on Pb accumulation in roots and leaves of $Pb(NO_3)_2$-treated *Zea mays* plants. Inset shows relative changes in Pb concentration. Amendment rate is given as *v/v* % added to soil. Data are means from 5 replicates for leaves and 3 replicates for roots ± SE. Different letters indicate statistically significant differences between the treatments according to Tukey's HSD test ($p < 0.05$). DM, dry mass. Control level of Pb was 1.34 and 3.01 mg kg^{-1} in leaves and roots, respectively.

4. Discussion

Lead tolerance mechanisms in plants are largely related to limiting its intake in roots and restricting its transport to above-ground parts. Root-released uronic-acid-containing exudates bind lead ions to their carboxyl groups, thus inhibiting their uptake [28]. The majority of lead taken up in roots is stored in cell walls in the form of relatively stable complexes with galacturonic and glucuronic acids [29]. A layer of thickened cells in endoderm tissue, called a Casparian strip, acts as a physical barrier for apoplastic transport of lead, which further limits its translocation to above-ground parts [30]. As a result, lead predominantly accumulates in the root tissues of plants. However, lead transport and accumulation in above-ground organs can be stimulated by chelating substances. In a classical study of Huang and Cunningham (1996) [31], *Z. mays* plants exhibited the highest shoot Pb accumulation potential among 11 plant species grown in Pb-contaminated soil, reaching 225 mg kg^{-1}. By use of chelating agents, it is possible to achieve a many-fold higher Pb accumulation in the shoots of *Z. mays*, reaching 771 mg kg^{-1} [32]. As an extreme, 10-day-old *Z. mays* seedlings cultivated in soil with 2500 mg kg^{-1} of Pb for 7 days accumulated more than 10,000 mg kg^{-1} of Pb in the shoots under the effect of synthetic chelate [31]. Due to this high accumulation potential and relatively high tolerance against Pb, *Z. mays* has been used as a model in phytoremediation studies [16].

In addition to genetic and physiological mechanisms, soil properties can significantly affect Pb availability for plants. Therefore, in the context of food security, agricultural practices leading to decreased heavy metal accumulation in *Z. mays* plants need to be considered. One such approach involves increasing the stabilization of heavy metals in soil, while providing sufficient mineral nutrient supply in plant-available forms in soil, as well as increasing soil sustainability. In this respect, organic fertilizers have drawn the greatest attention of researchers. Vermicompost is an organic fertilizer with a relatively high amount of plant-available nutrients and plant-growth-promoting substances, and has high microbiological activity, which make it a promising choice for soil remediation [7].

Usually, raw organic materials have less heavy metal absorption capacity in comparison to processed ones. Thus, the adsorption capacity for Pb of cow manure was observed to be 103 mg g^{-1}, compared to 171 mg g^{-1} of cow manure vermicompost [9]. In contrast, both sheep manure and vermicompost were ineffective as stabilizing materials of Pb, but biochars produced from the two materials were effective [10]. Highly variable effects of different organic materials on Pb accumulation in *Z. mays* and other plant tissues have been reported. From the opposite side, some studies used vermicompost as a material for the possibly stimulated accumulation of heavy metals in plant tissues [33,34]. In particular, a study with *Avena strigosa* indicated that the high rate of vermicompost addition to soil (>50%) increased the bioavailability of Cr and Pb, resulting in an enhanced accumulation of the metals in plant tissues [33]. However, this effect should not be confused with similar effects resulting from plant cultivation in the presence of sewage-sludge-derived vermicompost, as in the case of *Z. mays*, where vermicompost samples contained elevated concentrations of various heavy metals [14], and in the case of experiments with tannery sludge vermicompost [35].

Soil amendment with date palm leaf waste biochar (0.5–3%) decreased the soil availability of heavy metals and did not result in a change in Pb concentration in leaves of *Z. mays* plants, and there was a tendency of decreased root Pb accumulation by 23% at the highest amendment rate [36]. Organic formulation panchakavya decreased Pb accumulation in the shoots and roots of *Z. mays* by 32 and 37%, respectively [37]. The addition of chicken manure decreased Pb accumulation in *Z. mays* by 53% [38]. Vermicompost application (at 10% amendment rate) decreased Pb concentration in the shoots of *Brassica chinensis* by 67% [13]. Vermicompost decreased the soil plant-available concentration of Pb by 43% [12]. When maize straw biochar and maize straw compost was compared with respect to their ability to reduce heavy metal accumulation in *Z. mays* plants, the greatest effect was shown by biochar [39]. In contrast, while biochar had high Cd retention capacity, it also increased the Cd bioavailability and accumulation potential in plants [40].

The ability of organic materials to adsorb heavy metals, including Pb, has been mostly associated with the presence of humic substances [41,42]. However, the existence of other types of interaction cannot be ruled out, as humin was shown to be inefficient in decreasing the availability of Cd, in comparison to vermicompost and vermicompost solid residue [40]. The high concentration of Ca in the vermicompost sample used in the current study (Table 2) might have reduced the root uptake of lead, due to the inhibition of ion pumps in root cell membranes [43]. In addition, the high concentration of Fe^{2+} in the vermicompost might have negatively affected Pb uptake due to the antagonistic interaction between the two metals [44].

It has also been documented that the improvement of mineral supply promotes heavy metal accumulation in plants. The increased concentration of soil-available nitrogen stimulated Pb accumulation in *Z. mays* and *Spinacia oleraceae* plants [45]. This clearly was not the case in the present study, where the plant-available mineral nutrient pool in soil was enhanced by vermicompost amendment on the background of an increasing concentration of organic matter.

Differences in experimental conditions seem to greatly affect the results in this type of studies. One crucial aspect is related to the concentration of Pb (or any other heavy metal of interest) in soil samples used for plant cultivation in metal uptake and accumulation experiments. In some studies, the Pb concentration range was only several tens of mg kg^{-1} [39,46], while other studies used extremely high Pb loads (1000–2500 mg kg^{-1}) [31,34].

One of the possible reasons for Pb toxicity on plant physiological processes is associated with its negative effect on mineral nutrition, especially at the level of mineral element uptake [47]. The Pb treatment of *Z. mays* plants resulted in a significant reduction in K^+ concentration in the roots but not in the shoots [48]. The hypothesis that the shoot growth inhibition of *Z. mays* seedlings is due to its negative effect on the K^+ pool through K^+ leakage from the root cells has been tested but not confirmed [48]. Nitrogen metabolism at the level of nitrate represents another nutritional target of Pb toxicity [6]. This effect was clearly seen in the present study, as there was a decrease in nitrate accumulation in *Z. mays* even on the background of the additional nutrient supply through vermicompost amendment.

Leaf chlorophyll concentration and chlorophyll *a* fluorescence parameter Performance Index Total were not negatively affected by Pb treatment (Figures 7 and 8), while it is generally accepted that the decrease in chlorophyll concentration is one of the reasons for the diminished photosynthetic activity together with the negative effects on photosynthetic electron transport [49,50]. On the other hand, the increasing rate of vermicompost amendment resulted in a typical response of increased leaf chlorophyll concentration and Performance Index Total, showing an optimization of the physiological status of plants under the effect of vermicompost. A similar effect was observed previously in studies with different crops [51,52], and it is evident that leaf chlorophyll concentration positively responds to vermicompost amendment in a concentration-dependent manner. Most likely, this effect is associated with a prolonged growth period of leaves due to the better mineral supply of vermicompost-treated plants. However, the positive response of Performance Index Total to vermicompost amendment seems to be associated with the stimulation of activity of a water-splitting complex or other photochemical reactions at the donor side of photosystem II [52]. These responses are characteristic of plants at optimal mineral nutrient availability [22,23], reflecting the highest possible physiological performance of the plant.

In contrast to the effects on plant mineral nutrition, there were no significant negative morphological effects of Pb treatment on *Z. mays* plants, even for plants growing in soil without vermicompost amendment, while the total biomass of Pb-treated plants insignificantly decreased by 10% (Figure 2). It is possible that any negative effect of Pb was masked by the presence of an increased concentration of N in the substrate due to treatment with $PbNO_3$. When exposure to Pb nitrate and Pb acetate were compared using other model plants, it appeared that the two salts indeed had different effects [53–55]. When Pb acetate was used as a treatment in an experimental small plot field study with *Z. mays* plants, it significantly decreased the growth, morphological parameters, and grain yield [56].

When using high doses of Pb in the form of nitrate, as in the present study, it is impossible to dissect the effects of Pb from those of elevated nitrogen availability, without the use of an appropriate control. Especially when Pb-tolerant plants are used as model species, a seemingly positive effect of Pb on plant growth and physiological status can be due to the effect of surplus nitrogen [24,25]. In the present study, by using an additional control with the same amount of nitrogen as received in the Pb nitrate treatment, it was possible to dissociate Pb-specific effects from those of nitrate, both at morphological and physiological levels. Most importantly, Pb had a clear negative effect on NO_3 concentration in the base leaf, stem base, and roots, in comparison to the stimulative effect of surplus nitrogen (Figure 9), and a negative effect on K^+ concentration in the flag leaf, stem base, and roots (Figure 10). Interestingly, the negative effects of Pb treatment on plant height and biomass accumulation were efficiently prevented for plants grown in vermicompost-amended soil, due to a significant decrease in Pb accumulation capacity in plant tissues, but the above-mentioned negative effects on mineral nutrition were only diminished but not fully lost. The relationship between nitrogen and Pb was also affected not only by the presence of vermicompost, but also by the vermicompost amendment rate, as nitrogen treatment tended to give a higher stem and root biomass in comparison to Pb nitrate treatment at the 20% amendment rate, but the dry mass of leaves, stems, and roots was significantly lower in nitrogen-treated plants in comparison to the Pb-nitrate-treated plants at the 30% amendment rate (Figure 4). The total biomass of ammonium-nitrate-treated plants cultivated at the 30% vermicompost amendment rate was significantly decreased in comparison to that of the control and Pb-nitrate-treated plants (Figure 3). This more likely indicates the appearance of ammonium toxicity in conditions of high nutrient availability [57]. Recently, transcriptional signatures in roots of *Z. mays* have been compared for nitrate and ammonium, and it was shown that both overlapping and distinct pathways indeed are regulated [58].

Additional experiments in field conditions using natural contaminated soil and different forms of organic amendment at various rates are necessary for obtaining practically useful results, as the experimental system used had typical limitations characteristic for vegetation pot studies [59].

5. Conclusions

The main conclusion from this study was that, in addition to the pronounced positive effects of vermicompost soil amendment on the growth and physiology of *Z. mays* plants, it also significantly decreased Pb accumulation in plant leaves and roots. The most favorable effect was evident at 10 and 20% vermicompost amendment rates, resulting in a 65% decrease in Pb concentration in tissues of Pb-treated plants, while plant biomass increased four to five times in comparison to soil-grown control plants, together with accelerated flowering. Thus, vermicompost is one of most favorable and sustainable organic products for reducing heavy metal uptake and accumulation in crop plants, while also being an efficient organic fertilizer.

Author Contributions: Conceptualization, M.V. and G.I.; methodology, M.V.; investigation, Z.L.-S.; writing—original draft preparation, G.I. and Z.L.-S.; writing—review and editing, M.V. All authors have read and agreed to the published version of the manuscript.

Funding: This research received no external funding.

Institutional Review Board Statement: Not applicable.

Data Availability Statement: All data reported here are available from the authors upon request.

Acknowledgments: We thank Jānis Freibergs (Eko Zeme, Latvia) for providing the vermicompost sample used in the present study. We thank Guntis Brūmelis (University of Latvia) for English correction.

Conflicts of Interest: The authors declare no conflict of interest.

References

1. Pozza, L.E.; Field, D.J. The science of Soil Security and Food Security. *Soil Secur.* **2020**, *1*, 100002. [CrossRef]
2. Vidar, M. Soil and agriculture governance and food security. *Soil Secur.* **2022**, *6*, 100027. [CrossRef]
3. Rai, K.K.; Pandey, N.; Meena, R.P.; Rai, S.P. Biotechnological strategies for enhancing heavy metal tolerance in neglected and underutilized legume crops: A comprehensive review. *Ecotoxicol. Environ. Saf.* **2021**, *208*, 111750. [CrossRef] [PubMed]
4. Rai, P.K.; Lee, S.S.; Zhang, M.; Tsang, Y.F.; Kim, K.-H. Heavy metals in food crops: Health risks, fate, mechanisms, and management. *Environ. Int.* **2019**, *125*, 365–385. [CrossRef] [PubMed]
5. Kumar, A.; Kumar, A.; Cabral-Pinto, M.M.S.; Chaturvedi, A.K.; Shabnam, A.A.; Subrahmanyam, G.; Mondal, R.; Gupta, D.K.; Malyan, S.K.; Kumar, S.S.; et al. Lead toxicity: Health hazards, influence on food chain, and sustainable remediation approaches. *Int. J. Environ. Res. Public Health* **2020**, *17*, 2179. [CrossRef]
6. Pourrut, B.; Shahid, M.; Dumat, C.; Winterton, P.; Pinelli, E. Lead uptake, toxicity, and detoxification in plants. *Rev. Environ. Contam. Toxicol.* **2011**, *213*, 113–136. [PubMed]
7. Ievinsh, G. Review on physiological effects of vermicompost on plants. In *Biology of Composts*; Meghvansi, M.K., Varma, A., Eds.; Springer Nature Switzerland AG: Cham, Switzerland, 2020; pp. 63–86.
8. Matos, G.D.; Arruda, M.A.Z. Vermicompost as natural adsorbent for removing metal ions from laboratory effluents. *Process Biochem.* **2003**, *39*, 81–88. [CrossRef]
9. Zhu, W.; Du, W.; Shen, X.; Zhang, H.; Ding, Y. Comparative adsorption of Pb^{2+} and Cd^{2+} by cow manure and its vermicompost. *Environ. Pollut.* **2017**, *227*, 89–97. [CrossRef]
10. Boostani, H.R.; Najafi-Ghiri, M.; Hardie, A.G.; Khalili, D. Comparison of Pb stabilization in a contaminated calcareous soil by application of vermicompost and sheep manure and their biochars produced at two temperatures. *Appl. Geochem.* **2019**, *102*, 121–128. [CrossRef]
11. Zhang, Y.; Tian, Y.; Hu, D.; Fan, J.; Shen, M.; Zeng, G. Is vermicompost the possible in situ sorbent? Immobilization of Pb, Cd and Cr in sediment with sludge derived vermicompost, a column study. *J. Hazard. Mater.* **2019**, *367*, 83–90. [CrossRef]
12. Moslehi, A.; Feizian, M.; Higueras, P.; Eisvand, H.R. Assessment of EDDS and vermicompost for the phytoextraction of Cd and Pb by sunflower (*Helianthus annuus* L.). *Int. J. Phytoremed.* **2019**, *21*, 191–199. [CrossRef] [PubMed]
13. Wang, F.; Miao, L.; Wang, Y.; Zhang, M.; Zhang, H.; Ding, Y.; Zhu, W. Using cow dung and mineral vermireactors to produce vermicompost for use as a soil amendment to slow Pb^{2+} migration. *Appl. Soil Ecol.* **2022**, *170*, 104299. [CrossRef]
14. Zuo, W.; Xu, K.; Zhang, W.; Wang, Y.; Gu, C.; Bai, Y.; Shan, Y.; Dai, Q. Heavy metal distribution and uptake by maize in a mudflat soil amended by vermicompost derived from sewage sludge. *Environ. Sci. Pollut. Res.* **2019**, *26*, 30154–30166. [CrossRef] [PubMed]
15. Mwilola, P.N.; Mukumbuta, I.; Shitumbanuma, V.; Chishala, B.H.; Uchida, Y.; Nakata, H.; Nakayama, S.; Ishizuka, M. Lead, zinc and cadmium accumulation, and associated health risks, in maize grown near the Kabawe mine in Zambia in response to organic and inorganic soil amendments. *Int. J. Environ. Res. Public Health* **2020**, *17*, 9038. [CrossRef]
16. Wuana, R.A.; Okieimen, F.E. Phytoremediation potential of maize (*Zea mays* L.). A review. *Afr. J. Gen. Agric.* **2010**, *6*, 275–287.
17. Abedi, T.; Gavanji, S.; Mojiri, A. Lead and zinc uptake and toxicity in maize and their management. *Plants* **2022**, *11*, 1922. [CrossRef]
18. Cruz, Y.; Villar, S.; Gutiérrez, K.; Montoya-Ruiz, C.; Gallego, J.L.; Delgado, M.P.; Saldarriaga, J.F. Gene expression and morphological responses of *Lolium perenne* L. exposed to cadmium (Cd^{2+}) and mercury (Hg^{2+}). *Sci. Rep.* **2021**, *11*, 11257. [CrossRef] [PubMed]
19. Emamverdian, A.; Ding, Y.; Mokhberdoran, F.; Xie, Y. Heavy metal stress and some mechanisms of plant defense responses. *Sci. World J.* **2015**, *2015*, 756120. [CrossRef]
20. Paunov, M.; Koleva, L.; Vassilev, A.; Vangronsveld, J.; Goltsev, V. Effects of different metals on photosynthesis: Cadmium and zinc affect chlorophyll fluorescence in durum wheat. *Int. J. Mol. Sci.* **2018**, *19*, 787. [CrossRef]
21. Matisons, R.; Krišāns, O.; Jansons, Ā.; Kondratovičs, T.; Elferts, D.; Ievinsh, G. Norway spruce seedlings from an Eastern Baltic provenance show tolerance to simulated drought. *Forests* **2021**, *12*, 82. [CrossRef]
22. Wang, H.; Liu, R.L.; Jin, J.Y. Effects of zinc and soil moisture on photosynthetic rate and chlorophyll fluorescence parameters of maize. *Biol. Plant.* **2009**, *53*, 191–194. [CrossRef]
23. Horaczek, T.; Dąbrowski, P.; Kalaji, H.M.; Baczewska-Dąbrowska, A.H.; Pietkiewicz, S.; Stępień, W.; Gozdowski, D. JIP-test as a tool for early detection of the macronutrients deficiency in *Miscanthus* plants. *Phtosynthetica* **2020**, *58*, 507–517. [CrossRef]
24. Kovačević, B.; Miladinović, D.; Orlović, S.; Katanić, M.; Kebert, M.; Kovinčić, J. Lead tolerance and accumulation in white poplar cultivated in vitro. *South-east Eur. For. SEEFOR* **2013**, *4*, 3–12. [CrossRef]
25. Parys, E.; Wasilewska, W.; Siedlecka, M.; Zienkiewicz, M.; Drożak, A.; Romanowska, E. Metabolic responses to lead of metallicolous and nonmetallicolous populations of *Armeria maritima*. *Arch. Environ. Contam. Toxicol.* **2014**, *67*, 565–577. [CrossRef] [PubMed]
26. Osvalde, A. Optimization of plant mineral nutrition revisited: The roles of plant requirements, nutrient interactions, and soil properties in fertilization management. *Environ. Exp. Biol.* **2011**, *9*, 1–8.

27. Banks, J.M. Continuous excitation chlorophyll fluorescence parameters: A review for practicioners. *Tree Physiol.* **2017**, *37*, 1128–1136. [CrossRef]
28. Sharma, P.; Shanker, R. Lead toxicity in plants. *Braz. J. Plant Physiol.* **2005**, *17*, 35–52. [CrossRef]
29. Inoue, H.; Fukuoka, D.; Tatai, Y.; Kamachi, H.; Hayatsu, M.; Ono, M.; Suzuki, S. Properties of lead deposits in cell walls of radish (*Raphanus sativus*) roots. *J. Plant Res.* **2013**, *126*, 51–61. [CrossRef]
30. Seregin, I.V.; Shpigun, L.K.; Ivanov, V.B. Distribution and toxic effects of cadmium and lead on maize roots. *Russ. J. Plant Physiol.* **2004**, *51*, 525–533. [CrossRef]
31. Huang, J.W.; Cunningham, D. Lead phytoextraction: Species variation in lead uptake and translocation. *New Phytol.* **1996**, *134*, 75–84. [CrossRef]
32. Li, H.; Wang, Q.; Cui, Y.; Dong, Y.; Christie, P. Slow release chelate enhancement of lead phytoextraction by corn (*Zea mays* L.) for contaminated soil—a preliminary study. *Sci. Total Environ.* **2005**, *339*, 179–187. [CrossRef] [PubMed]
33. Hoehne, L.; de Lima, C.V.S.; Martini, M.C.; Altmayer, T.; Brietzke, D.T.; Finatto, J.; Gonçalves, T.E.; Granada, C.E. Addition of vermicompost to heavy metal-contaminated soil increases the ability of black oat (*Avena strigosa* Schreb) plants to remove Cd, Cr, and Pb. *Water Air Soil Pollut.* **2016**, *227*, 443. [CrossRef]
34. Kouhi, S.M.M.; Ardakani, M.R.S.; Beykkhormizi, A. Irrigation of *Helianthus annuus* with Pb-polluted water: Improvement of phytoremediation using vermicompost. *J. Plant Process Funct.* **2019**, *8*, 77–83.
35. Gondek, K.; Filipek-Maazur, B. Biomass yields of shoots and roots of plants cultivated in soil amended by vermicomposts based on tannery sludge and content of heavy metals in plant tissues. *Plant Soil Environ.* **2003**, *49*, 402–409. [CrossRef]
36. Ahmad, M.; Usman, A.R.A.; Al-Faraj, A.S.; Ahmad, M.; Sallam, A.; Al-Wabel, M.I. Phosphorus-loaded biochar changes soil heavy metals availability and uptake potential of maize (*Zea mays* L.) plants. *Chemosphere* **2018**, *194*, 327–339. [CrossRef] [PubMed]
37. Praburaman, L.; Park, J.-H.; Park, Y.-J.; He, Z.; Kamala-Kannan, S.; Oh, B.-T. Effect of panchakavya (organic formulation) on phytoremediation of lead and zinc using *Zea mays*. *Chemosphere* **2020**, *246*, 125810. [CrossRef]
38. Guo, X.; Han, W.; Zhang, G.; Yang, W.; Wei, Z.; He, Q.; Wu, Q. Effect of inorganic and organic amendments on maize biomass, heavy metals uptake and their availability in calcareous and acid washed soil. *Environ. Technol. Innov.* **2020**, *19*, 101038. [CrossRef]
39. Irfan, M.; Mudassir, M.; Khan, M.J.; Dawar, K.M.; Muhammad, D.; Mian, I.A.; Ali, W.; Fahad, S.; Saud, S.; Hayat, Z.; et al. Heavy metals immobilization and improvement of maize (*Zea mays* L.) growth amended with biochar and compost. *Sci. Rep.* **2021**, *11*, 18416. [CrossRef]
40. Pinto, T.O.; García, A.C.; Guedes, J.N.; Sobrinho, N.M.B.A.; Tavares, O.C.H.; Berbara, R.L.L. Assessment of the use of natural materials for the remediation of cadmium soil contamination. *PLoS ONE* **2016**, *11*, e0157547.
41. Trevisan, S.; Francioso, O.; Quaggiotti, S.; Nardi, S. Humic substances biological activity at the plant-soil interface. *Plant Signal. Behav.* **2010**, *5*, 635–643. [CrossRef]
42. Meng, F.; Yuan, G.; Wei, J.; Bi, D.; Ok, Y.S.; Wang, H. Humic substances as a washing agent for Cd-contaminated soils. *Chemosphere* **2017**, *181*, 461–467. [CrossRef] [PubMed]
43. Kozhevnikova, A.D.; Seregin, I.V.; Bystrova, E.I.; Belyaeva, A.I.; Kataeva, M.N.; Ivanov, V.B. The effects of lead, nickel, and strontium nitrates on cell division and elongation in maize roots. *Russ. J. Plant Physiol.* **2009**, *56*, 242–250. [CrossRef]
44. Dong, Y.; Ma, L.Q.; Rhue, R.D. Relation of enhanced Pb solubility to Fe partitioning in soils. *Environ. Pollut.* **1999**, *110*, 515–522. [CrossRef] [PubMed]
45. Peñafiel-Sandoval, Z.B.; Iannacone, J. Effect of urea on lead absorption in corn (*Zea mays* L.), spinach (*Spinacia oleraceae* L.) and cabbage (*Brassica oleraceae* L.). *Agron. Columb.* **2020**, *38*, 105–217.
46. Kacálková, L.; Tlustoš, P.; Száková, J. Chromium, nickel cadmium, and lead accumulation in maze, sunflower, willow, and poplar. *Pol. J. Environ. Stud.* **2014**, *23*, 753–761.
47. Hadi, F.; Aziz, T. A mini review on lead (Pb) toxicity in plants. *J. Biol. Life Sci.* **2015**, *6*, 91–101. [CrossRef]
48. Małkowski, E.; Kita, A.; Galas, W.; Karcz, W.; Kuperberg, J.M. Lead distribution in corn seedlings (*Zea mays* L.) and its effect on growth and the concentrations of potassium and calcium. *Plant Growth Regul.* **2002**, *37*, 69–76. [CrossRef]
49. Sengar, R.S.; Pandey, M. Inhibition of chlorophyll biosynthesis by lead in greening *Pisum sativum* leaf segments. *Biol. Plant.* **1996**, *38*, 459–462. [CrossRef]
50. Kumar, A.; Prasad, M.N.V. Lead-induced toxicity and interference in chlorophyll fluorescence in *Talinum triangulare* grown hydroponically. *Photosynthetica* **2015**, *53*, 66–71. [CrossRef]
51. Ievinsh, G.; Andersone-Ozola, U.; Zeipiņa, S. Comparison of the effects of compost and vermicompost soil amendments in organic production of four herb species. *Biol. Agric. Hortic.* **2020**, *36*, 267–282. [CrossRef]
52. Ose, A.; Andersone-Ozola, U.; Ievinsh, G. Substrate-dependent effect of vermicompost on yield and physiological indices of container-grown *Dracocephalum moldavica* plants. *Agriculture* **2021**, *11*, 1231. [CrossRef]
53. Sharma, S.D.; Chorpa, R.N. Effect of lead nitrate and lead acetate on growth of the moss *Semibarbula orientalis* (Web.) Wijk. et Marg. grown in vitro. *J. Plant Physiol.* **1987**, *129*, 243–249. [CrossRef]
54. Yang, N.; Zhou, F.-R.; WEang, J.-X. Eco-toxicological effects of two kinds of lead compounds on forest tree seed in alkaline soil. *Environ. Monit. Assess.* **2016**, *188*, 201. [CrossRef] [PubMed]
55. Alexandrino, R.C.S.; Lima, F.R.D.; Martins, G.C.; Natal-du-Luz, T.; Sousa, J.P.; Guilherme, L.R.G.; Marques, J.J. Lead acetate toxicity in tropical soils. *Ecotoxicology* **2021**, *30*, 1029–1042. [CrossRef] [PubMed]

56. Sofy, M.R.; Seleiman, M.F.; Alhammad, B.A.; Alharbi, B.M.; Mohamed, H.I. Minimizing adverse effects of Pb on maize plants by combined treatment with jasmonic, salicylic acids and proline. *Agronomy* **2020**, *10*, 699. [CrossRef]
57. Britto, D.T.; Kronzucker, H.J. NH_4 toxicity in higher plants: A critical review. *J. Plant Physiol.* **2002**, *159*, 567–584. [CrossRef]
58. Ravazollo, L.; Trevisan, S.; Forestan, C.; Varotto, S.; Sut, S.; Dall'Acqua, S.; Malagoli, M.; Quaggiotti, S. Nitrate and ammonium affect the overall maize response to nitrogen availability by triggering specific and common transcriptional signatures in roots. *Int. J. Mol. Sci.* **2020**, *21*, 686. [CrossRef]
59. Poorter, H.; Fiorani, F.; Stitt, M.; Schurr, U.; Finck, A.; Gibon, Y.; Usadel, B.; Munns, R.; Atkin, O.K.; Tardieu, F.; et al. The art of growing plants for experimental purposes: A practical guide for the plant biologist. *Funct. Plant Biol.* **2012**, *39*, 821–838. [CrossRef]

agriculture

MDPI

Article

How the Composition of Substrates for Seedling Production Affects Earthworm Behavior

Sandra Ečimović [1], Mirna Velki [1,*], Alma Mikuška [1], Jelena Bažon [1], Lucija Sara Kovačić [1], Suzana Kristek [2], Jurica Jović [2], Franjo Nemet [2], Katarina Perić [2] and Zdenko Lončarić [2]

[1] Department of Biology, Josip Juraj Strossmayer University of Osijek, Cara Hadrijana 8/A, 31000 Osijek, Croatia

[2] Faculty of Agrobiotechnical Sciences Osijek, Josip Juraj Strossmayer University of Osijek, Vladimira Preloga 1, 31000 Osijek, Croatia

* Correspondence: mirna.velki@gmail.com or mvelki@biologija.unios.hr

Abstract: The constant increase in the intensity of agricultural production simultaneously increases the risk of negative effects of long-term agricultural practices. By-products of agricultural, forestry, and food production, as well as other types of organic waste, can be used as raw materials in the production of organic fertilizers and substrates for seedling cultivation through various processes of biological stabilization. In this way, the amount of waste is reduced, which contributes to the preservation of soil fertility and the sustainable use of resources. During waste processing and the stabilization of organic matter can be improved by using earthworms (vermicomposting). The aim of this study was to determine how different substrates, composed of different components and their mixtures, affect the earthworm *Eisenia andrei*. The effects of investigated substrates on the survival and behavior of earthworms were monitored. In addition, the effect of tested substrates on acetylcholinesterase (AChE), carboxylesterase (CES), and glutathione S-transferase (GST) activity was also assessed. The results showed that the most suitable substrates were leaves with horse manure and grape pomace alone and in combination with rock wool and sawdust. The obtained results provide important information on components and mixtures that have the greatest potential in the production of organic fertilizers and substrates for growing seedlings.

Keywords: organic waste; organic fertilizers; earthworms; vermicomposting; seedlings

1. Introduction

The need for agricultural production is constantly growing and, at the same time, increasing the risks related to the negative effects of long-term agricultural practices, such as the usage of mineral fertilizers and pesticides. Due to the possible accumulation of these agrochemicals and residues in ecosystems and food chains, these agricultural practices may cause adverse effects on the health of animals and humans. Therefore, it is necessary to find alternatives to the usage of agrochemicals with the aim to develop a sustainable plant production system without the usage of mineral fertilizers [1,2]. The first step in this process is to determine the potential material that can be used for such a purpose. One of the options is to use by-products of agriculture, forestry, and food production, as well as various types of organic waste, as raw materials for the production of fertilizers and substrates. On one hand, these by-products and waste represent a problem since they have to be removed, but on the other hand, they are a potential reservoir of organic matter that can be usefully exploited.

Organic waste has the potential for application in the production of substrates that can be used for seedling cultivation. Different types of biowaste can be used as components in substrates and, considering their characteristics, can be mixed in different ratios in order to obtain a product of specific features. For example, animal manure, aquaculture sludge, grape pomace, rock wool, sawdust, wood chips, leaves, vegetable market waste, rice straw, etc., can be used [3–7]. The management of organic waste is extremely important since it provides several benefits, such as reducing the amount of waste, decreasing the production of certain toxic gases, leaching, which can cause environmental pollution, and obtaining high value products for the production of fertilizers [8,9].

Given that resources should be used sustainably and attention should be focused on organic production in these processes, vermicomposting plays an important role. Vermicomposting is the eco-biotechnological process of the decomposition of organic matter through earthworms and their symbiotic microorganisms' activities [10,11]. As a result of this process, earthworm castings (vermicompost) are obtained. Vermicompost is a high-quality organic fertilizer that is rich in microbial activity and plant growth regulators and plant nutrients [12]. Vermicompost application can be helpful in the improvement of soil biophysical, chemical properties, and fertility, as well as in the remediation of the soil [7,13,14]. Earthworms are one of the most important animal communities in soil ecosystems. They have the capability to cultivate and aerate the soil, stimulate microbial activity, and contribute significantly to the organic matter decomposition into nutrient-rich products by improving soil properties [15]. In addition, they can act to reduce pathogens, since they possess different antimicrobial and antifungal molecules [16,17]. The digestive system of earthworms has the ability to transform various materials, such as plant, animal, industrial, and urban wastes, into beneficial vermicompost [18,19]. In the vermicomposting process, the *Eisenia foetida* and *Lumbricus* spp. species are the most preferred species due to the following characteristics. They have a short life cycle, a high reproductive rate, a low body weight, and a high resistance and tolerate variable ranges of temperatures [20,21]. Vermicomposting can also indirectly mitigate the effects of global warming and the greenhouse effect through its ability to sequester carbon in the soil [22]. The more organic carbon is retained in the soil, the more the extenuation potential of agriculture against climate change is higher. Unlike traditionally produced compost, vermicompost is characterized by excellent structure, high porosity, drainage, and water-holding capacity. It contains many useful macro and micronutrients in plant-available forms (nitrogen, phosphorus, potassium, organic carbon, calcium, sulfur, magnesium, hormones, vitamins, and enzymes) and it is an excellent alternative to mineral fertilizers [10,12,23]. Additionally, vermicompost provides many other plant-useful components such as vitamins, hormones, humic substances, and antioxidants [24]. Vermicomposting is a completely natural way of obtaining quality fertilizer which is important for soil fertility, as well as for the growth of many plants and agricultural crops [25].

Considering the potential for the usage of organic waste and the benefits resulting from the decomposition of organic matter through vermicomposting, it is important to determine which type of organic waste could be used in such a process. Even though different substrates have already been used, there are many components that could be used in different combinations and, consequently, there is a lack of information on the potentially suitable mixtures. Therefore, in the present study, different organic, as well as inorganic, components were selected for assessment and mixed in different combinations and ratios. In addition to new knowledge on the suitability of such substrates, an important aspect is also the reduction in waste that can be achieved with the usage of these substrates. This study will provide important information on the production of organic fertilizers and substrates for growing seedlings and support the principles of sustainable agriculture.

The main aim was to investigate which substrates, based on their composition, could be subjected to the vermicomposting process in order to enhance their characteristics and make them suitable for the production of organic fertilizers and substrates for growing seedlings. In order to determine that, the following aspects were examined: (1) the determination of how different substrates affect the survival of the earthworm *Eisenia andrei*; (2) the assessment of the behavior of the earthworm *Eisenia andrei* in different substrates in terms of the determination of the potential preference or avoidance of certain substrate or its components; and (3) the determination of effects of different substrates on activities of acetylcholinesterase (AChE), carboxylesterase (CES), and glutathione S-transferase

(GST). Substrates suitable for vermicomposting should not cause mortality, should not be avoided by earthworms, and should not cause changes in measured biochemical biomarkers. Based on the observed responses and comparisons between results, it will be possible to determine which substrates, or specific components of a substrate, have the greatest potential for further processing in the production of organic fertilizers and substrates for growing seedlings.

2. Materials and Methods

2.1. Test Organism

Exposures were conducted using adult specimens of the earthworm *Eisenia andrei* [26] (Oligochaeta, Lumbricidae). Earthworms were purchased from a local supplier and placed in cultivation containers in order to acclimatize prior to experiments (at least 2 weeks at 20 °C). Prior to usage in the experimental set-up, earthworms were separated and rinsed with tap water and stored in Petri dishes on damp filter paper for 12 h, in the dark at 20 ± 2 °C, allowing to empty the gut contents [27].

2.2. Substrates

In order to investigate the effects of different substrates, i.e., specific components of the substrates, on earthworm survival, behavior, and biomarker responses, three sets of substrates were prepared. Each set is comprised of six substrates and the details are given below (Tables 1–3). Additional parameters of investigated substrates (electrical conductivity, moisture, organic content, and C/N ratio) are given in Table S1. The components of the substrates were chosen based on their potential to be used in the production of organic fertilizers and substrates.

Table 1. Details of the first set of substrates.

Substrate Label	Substrate Composition	Composition Abbreviations	Ratio (*v*/*v*)	Number of Days in the Thermophilic Phase	pH
S1.1	Leaves and horse manure	a + b	1:1 (a:b)	8	9.11
S1.2	Leaves, horse manure, microorganisms, and urea	a + b + c * + d **	1:1 (a:b)	30	8.73
S1.3	Leaves, horse manure, microorganisms, wood chips, and phosphorite	a + b + c * + e + f ***	2.5:2.5:1 (a:b:e)	11	9.25
S1.4	Leaves, microorganisms, urea, and wood chips	a + c * + d ** + e	2.5:1 (a:e)	14	8.83
S1.5	Leaves and wood chips	a + e	2.5:1 (a:e)	12	8.83
S1.6	Leaves, urea, and wood chips	a + d ** + e	2.5:1 (a:e)	10+5	9.12

a—leaves; b—horse manure (straw was used as bedding); c—microorganisms; d—urea; e—wood chips; f—phosphorite. * The following microorganisms were added to S1.2: *Pseudomonas*, *Azotobacter*, *Azospirillum*, and *Bacillus* and, to S1.3 and S1.4, *Pseudomonas*, *Azotobacter*, and *Azospirillum*. Detailed information on the microorganisms is given in Table S2. ** Urea was added in the amount of 2.25 kg/m^3. *** Phosphorite was added in the amount of 2 kg/m^3.

Table 2. Details of the second set of substrates.

Substrate Label	Substrate Composition	Composition Abbreviations	Ratio (w/w)	Number of Days in the Thermophilic Phase	pH
S2.1	Grape pomace	g		-	7.33
S2.2	Fresh horse manure	h		-	7.71
S2.3	Grape pomace and fresh horse manure	g + h	1:1	-	7.14
S2.4	Grape pomace and rock wool	g + j	4:1	-	7.33
S2.5	Fresh horse manure and rock wool	h + j	4:1	-	8.23
S2.6	Grape pomace, fresh horse manure, and rock wool	g + h + j	2:2:1	-	7.78

g—grape pomace; h—fresh horse manure (sawdust was used as bedding); i—sawdust; j—rock wool.

Table 3. Details of the third set of substrates.

Substrate Label	Substrate Composition	Composition Abbreviations	Ratio (w/w)	Number of Days in the Thermophilic Phase	pH
S3.1	Fresh horse manure and sawdust	h + i	4:1	-	7.59
S3.2	Grape pomace and sawdust	g + i	4:1	-	6.42
S3.3	Grape pomace, sawdust, and rock wool	g + i + j	3:1:1	-	7.24
S3.4	Grape pomace, fresh horse manure, rock wool, and sawdust	g + h + i + j	1.5:1.5:1:1	-	7.59
S3.5	Grape pomace, fresh horse manure, and sawdust	g + h + i	2:2:1	-	6.69
S3.6	Fresh horse manure, sawdust, and rock wool	h + i + j	3:1:1	-	7.31

g—grape pomace; h—fresh horse manure (sawdust was used as bedding); i—sawdust; j—rock wool.

2.3. Acute Toxicity Test

In order to determine the survival of earthworms in a particular substrate, an acute toxicity test was conducted. The test was performed following standardized OECD guidelines [26] with some modifications, as required by the research objectives. Namely, since the aim was to determine the potential toxicity of the investigated substrates, instead of in the artificial soil, earthworms were placed in the test substrates. All exposures were performed in three replicates at 20 °C. The substrates (400 g) were placed into plastic containers (18 cm length, 16 cm width, and 10 cm height), followed by the addition of ten earthworms to each container. The containers were covered with a lid with ventilation holes. Survival rates were assessed after 48 h and 14 days. In each container, surviving earthworms were sorted by hand. Earthworms were considered dead when they did not respond to gentle touching.

2.4. Avoidance Tests with Earthworms

The avoidance experiments were carried out following standardized guidelines [28]. Two separate chambers were created in a plastic container (18 cm length, 16 cm width, and 10 cm height) divided by using a metal divider that was placed in the middle of the box. Each chamber was filled with 200 g of different substrates. After the addition of substrates on both sides of the container, the metal separator was removed, and ten adult earthworms were placed on the separating line. All exposures were performed in three replicates. The containers were covered with a lid with ventilation holes and incubated at

20 °C in the dark for 48 h. At the end of the test period, substrates were again separated by inserting the metal divider and the number of earthworms on each side was determined by hand sorting.

2.5. Biomarker Assessment

2.5.1. Chemicals

All chemicals used in the study were of analytical grade. The following chemicals were used: acetonitrile (C_2H_3N, CAS 75-05-8), 1-chloro-2,4-dinitrobenzene (CDNB) ($C_6H_3ClN_2O_4$, CAS 97-00-7), acetylthiocholine iodide ($CH_3COSCH_2CH_2N(CH_3)_3I$, CAS 1866-15-5), disodium hydrogen phosphate (NaH2PO4, CAS 7558-79-4), 5,5′-dithiobis-(2-nitrobenzoic acid) (DTNB) ($[-SC_6H_3(NO_2)CO_2H]_2$, CAS 69-78-3), 4-nitrophenyl acetate ($C_8H_7NO_4$, CAS 830-03-5), (2S)-2-amino-4-{[(1R)-1-[(carboxymethyl)carbamoyl]-2-sulfanylethyl]carbamoyl}butanoic acid (glutathione (GSH)) ($C_{10}H_{17}N_3O_6S$, CAS 70-18-8), and sodium dihydrogen phosphate dihydrate ($NaH_2PO_4 \times 2H_2O$, CAS 13472-35-0).

2.5.2. Sample Preparation

After the exposure period ended, earthworms were removed from the soil, thoroughly cleaned with distilled water, and dried. Earthworms were then individually homogenized on ice with the addition of a cold sodium phosphate buffer (0.1 M, pH 7.2, and a ratio of 1:5 *w:v*) with an Ultra-Turrax T10 homogenizer (IKA, Königswinter, Germany). The homogenates were centrifuged for 30 min at 9000× *g* at 4 °C to yield the post-mitochondrial fraction (supernatant S9). The aliquots of the S9 samples were stored at −80 °C until biomarker measurements.

2.5.3. Enzymatic Activities Evaluation

Protein content was measured according to the method first described by Smith et al. (1985) [29] using the PierceTM BCA Protein Assay Kit (Thermo Fisher Scientific, Waltham, MA, USA). The samples were diluted in a sodium phosphate buffer (0.1 M, pH 7.2), followed by the addition of a working solution. The absorbance was measured after a 2 h incubation period at 562 nm. The amount of proteins was calculated based on the calibration curve with BSA.

Acetylcholine esterase (AChE; 3.1.7) activity was assessed according to the method of Ellman et al. (1961) [30]. The reaction mixture included sodium phosphate buffer (0.1 M, pH 7.2), DTNB (1.6 mM), acetylcholine iodide (156 mM), and the sample (S9). Changes in absorbance were recorded during 2 min at 412 nm and the specific enzyme activity was expressed as nmol of acetylthiocholine iodide hydrolyzed in one min per mg of proteins.

Carboxylesterase (CES; EC 3.1.1.1.) activity was determined according to Hosokawa and Satoh (2002) [31]. The reaction mixture was comprised of 4-nitrophenylacetate and the sample (S9). Kinetics were recorded for 1 min at 405 nm and the specific enzyme activity was expressed as nmol of 4-nitrophenol produced per min per mg of protein.

Glutathione S-transferase (GST; EC 2.5.1.18.) activity was determined according to the method of Habig et al. (1974) [32]. The assay mixture consisted of CDNB (1mM), GSH (25 mM), and the sample (S9). Kinetics was recorded for 2 min at 340 nm and the specific enzyme activity was expressed as nmol of conjugated GSH in one min per mg of protein.

All measurements were performed in technical triplicates in 96-well microplates using a Tecan Spark microplate reader (Tecan Trading AG, Männedorf, Switzerland).

2.6. Data Analysis

Data obtained from experiments performed in three independent biological replicates have been analyzed. The percentage of survival was calculated based on the number of surviving earthworms in each substrate. Data on the behavior and biomarker responses were analyzed using GraphPad Prism 9 (GraphPad Software, San Diego, CA, USA). Data were first checked for homoscedasticity (the Bartlett test) and normality (the Shapiro–Wilk test). Since the normal distribution of the data was determined, parametrical tests were applied. For the avoidance behavior, significant differences in the preference of the earthworms for control or exposed soil were determined by means of Student's *t*-test. Biomarker data were analyzed with one-way ANOVA followed by Tukey's post hoc test. The probability level for statistical significance was set to $p < 0.05$ throughout the study.

3. Results

The first step in the experimental set-up was the investigation of the potential toxicity of tested substrates in terms of causing the mortality of earthworms. For that purpose, earthworms were placed in tested substrates for 48 h and 14 days, and the survival rate was assessed. Since the survival of earthworms is crucial for the determination of the suitability of substrate for vermicomposting, based on the mortality results following the experiments that were performed. Namely, only substrates in which earthworm survival rate was over 80% were used in the following experiments. In the case of lower survival, substrates were considered inadequate for further processing by earthworms. Further investigations included behavior assessment and measurement of biomarker responses.

3.1. Survival Rate

The survival rate results are presented in Table 4. The survival rate (expressed in %) was calculated from the number of earthworms that were found alive in the substrate. Missing earthworms, if any, were considered dead. Based on the survival rate results, substrates S1.2, S1.4, S2.2, S3.1, S3.4, S3.5, and S3.6 were excluded from further investigations.

Table 4. Survival rate (%) of the earthworms *Eisenia andrei* exposed to tested substrates for 48 h and 14 days.

Substrate	Survival Rate after 48 h	Survival Rate after 14 Days
S1.1	100%	100%
S1.2	0%	0%
S1.3	100%	100%
S1.4	0%	0%
S1.5	100%	100%
S1.6	100%	100%
S2.1	100%	95%
S2.2	30%	25%
S2.3	95%	85%
S2.4	90%	80%
S2.5	100%	90%
S2.6	95%	90%
S3.1	73%	50%
S3.2	100%	100%
S3.3	90%	90%
S3.4	53%	43%
S3.5	63%	53%
S3.6	57%	47%

Substrates labeled in grey were excluded from further assessments due to a low survival rate (<80%).

In the first set of substrates, where substrates S1.2 and S1.4 were excluded from further experiments, it seems that the addition of microorganisms and/or urea caused unfavorable conditions for earthworms and, consequently, led to mortality.

In the second set of substrates, only substrate S2.2 had to be excluded, indicating that fresh horse manure is not an adequate substrate for earthworms unless other components are added.

In the third set of substrates, four substrates (S3.1, S3.4, S3.5, and S3.6) caused high mortality and were consequently excluded from further investigations. Here again, fresh horse manure was the problematic component; however, the addition of other components was not adequate to neutralize fresh horse manure's adverse effects, indicating the importance of proper component selection.

3.2. *Behavior Assessment*

In order to determine whether earthworms have a preference for a particular substrate (due to different components of the substrate), the avoidance behavior of earthworms was assessed. Since three separate sets of substrates were prepared, avoidance behavior was assessed for each set separately.

3.2.1. First Set of Substrates

In the first set of substrates, after the mortality assessment, four (of six) substrates were tested for avoidance behavior. The results are presented in Table 5.

Table 5. Results of the avoidance test with the earthworms *Eisenia andrei* exposed to a first set of tested substrates for 48 h.

Distribution		Significance	Result
S1.1 90%	S1.3 10%	***	preference of substrate S1.1
S1.1 80%	S1.5 20%	**	preference of substrate S1.1
S1.1 67%	S1.6 33%	**	preference of substrate S1.1
S1.3 53%	S1.5 47%	NS	-
S1.3 48%	S1.6 52%	NS	-
S1.5 43%	S1.6 57%	NS	-

Significant differences between substrates (t-test) are labeled with ** ($p < 0.01$) and *** ($p < 0.001$). NS—not significant.

The results of the behavioral assessment showed that in the first set of tested substrates, earthworms preferred substrate S1.1, i.e., a combination of horse manure and leaves. Even though horse manure and leaves were also present in some of the other substrates, in this set this combination proved to be the most adequate for earthworms.

3.2.2. Second Set of Substrates

In the second set of substrates, after the mortality assessment, five (of six) substrates were tested for avoidance behavior. The results are presented in Table 6.

Table 6. Results of the avoidance test with the earthworms *Eisenia andrei* exposed to a second set of tested substrates for 48 h.

Distribution		Significance	Result
S2.1 93%	S2.3 7%	***	preference of substrate S2.1
S2.1 33%	S2.4 67%	NS	-
S2.1 100%	S2.5 0%	***	preference of substrate S2.1

Table 6. *Cont.*

Distribution		Significance	Result
S2.1 77%	S2.6 23%	**	preference of substrate S2.1
S2.3 10%	S2.4 90%	***	preference of substrate S2.4
S2.3 50%	S2.5 50%	NS	-
S2.3 3%	S2.6 97%		preference of substrate S2.6
S2.4 97%	S2.5 3%		preference of substrate S2.4
S2.4 97%	S2.6 3%		preference of substrate S2.4
S2.5 27%	S2.6 73%		preference of substrate S2.6

Significant differences between substrates (*t*-test) are labeled with ** ($p < 0.01$) and *** ($p < 0.001$). NS—not significant.

In the second set of substrates, the behavioral assessment showed preferences for substrates S2.1, S2.4, and S2.6. When analyzing the composition of these substrates, it is visible that all preferred substrates contained grape pomace, indicating its suitability as a substrate component.

3.2.3. Third Set of Substrates

In the third set of substrates, after the mortality assessment, two (of six) substrates were tested for avoidance behavior. The results are presented in Table 7.

Table 7. Results of the avoidance test with the earthworms *Eisenia andrei* exposed to a third set of tested substrates for 48 h.

Distribution		Significance	Result
S3.2 23%	S3.3 77%	***	preference of substrate S3.3

Significant differences between substrates (*t*-test) are labeled with *** ($p < 0.001$).

In the third set, only two substrates were tested and, considering their composition it, seems that the addition of rock wool to the substrate contributed to its favorable characteristics for the earthworms.

3.3. *Biomarker Responses*

Responses of selected biomarkers in earthworms, exposed to all substrates that were analyzed in behavioral assessment, were evaluated. Exposure of earthworms to substrates from the second and third sets did not result in significant changes, so those results are not shown. Biomarkers measured in earthworms placed in substrates from the first set showed some significant differences, and the results are presented in Figures 1–3.

Biomarker responses showed that in all three measured biomarkers, significant differences between substrates were determined. Already 48 h after exposure to tested substrates, the differences in measured activities were observed indicating that, in addition to the behavioral changes in terms of preferences for certain substrates, components of the substrates could have effects also on a biochemical level.

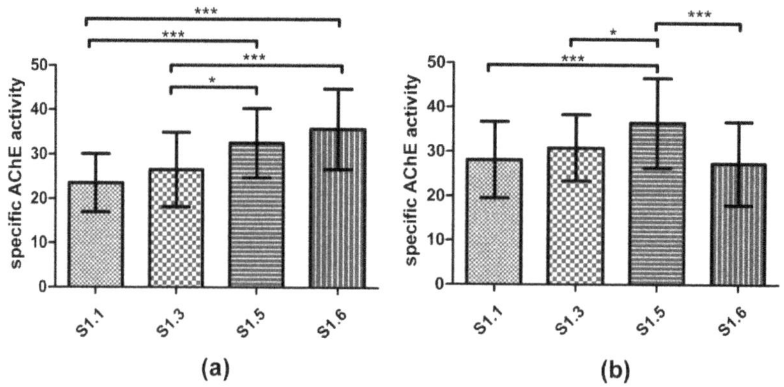

Figure 1. Specific acetylcholinesterase (AChE) activity was measured in earthworms exposed to the first set of substrates for 48 h (**a**) and 14 days (**b**). Significant differences between substrates (ANOVA followed by Tukey) are labeled with * ($p < 0.05$) and *** ($p < 0.001$).

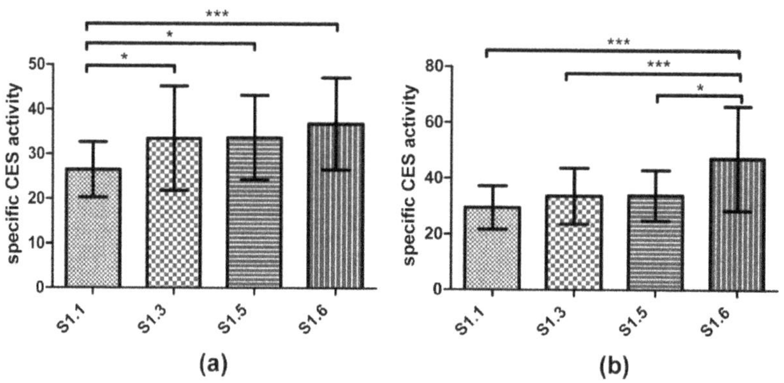

Figure 2. Specific carboxylesterase (CES) activity was measured in earthworms exposed to the first set of substrates for 48 h (**a**) and 14 days (**b**). Significant differences between substrates (ANOVA followed by Tukey) are labeled with * ($p < 0.05$) and *** ($p < 0.001$).

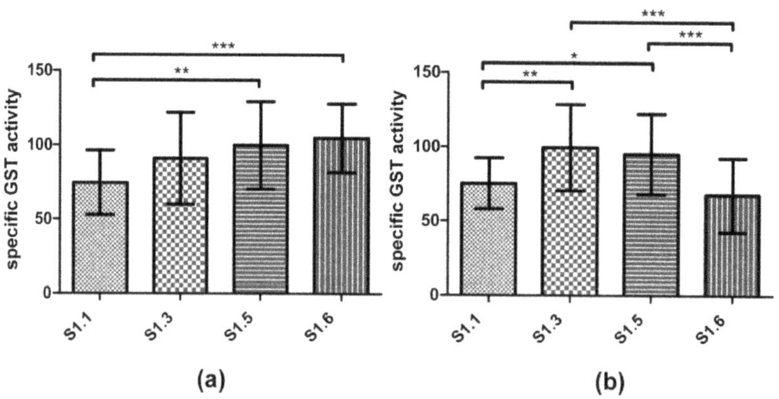

Figure 3. Specific glutathione S-transferase (GST) activity was measured in earthworms exposed to the first set of substrates for 48 h (**a**) and 14 days (**b**). Significant differences between substrates (ANOVA followed by Tukey) are labeled with * ($p < 0.05$), ** ($p<0.01$), and *** ($p < 0.001$).

4. Discussion

The success of vermicomposting is significantly influenced by the type of substrate added to the process [33]. In this sense, before the vermicomposting process, it is important to carry out certain tests with selected substrates. In this way, it is possible to choose components that will be most adequate for this process. Selection of the most adequate substrate will result in the optimal processing of material and obtaining a product of high value.

In the present study, three substrate sets, each comprising six substrates with different compositions, were tested to assess their effects on the earthworm *Eisenia andrei*. In the first set, the substrates were first subjected to composting (days spent in the thermophilic phase are given in Table 1). After that, earthworms were placed in the obtained substrates and two substrates proved to be completely unsuitable for vermicomposting, as substantial mortality was observed already after 48 h of exposure (substrates S1.2 and S1.4). In other substrates, survival was 100%, which enabled further behavioral and biomarker testing. Regarding the mortality in mixtures of leaves, horse manure, microorganisms and urea (S1.2), leaves, microorganisms, urea, and wood chips (S1.4), it is possible that the addition of microorganisms caused unfavorable conditions and, consequently, led to the mortality of earthworms. There is also a possibility that the urea was not sufficiently balanced in the mixture, which may be the cause of mortality of earthworms in these substrate combinations. The remaining substrates (S1.1, S1.3, S1.5, and S1.6) were further tested in an avoidance test. Namely, the avoidance test is based on the fact that organisms have the ability to avoid unfavorable conditions. This test is quick, cost-effective, ecologically relevant, and has a high sensitivity. Avoidance behavior by earthworms has been recognized as a valuable endpoint in soil quality assessment [34,35]. In the avoidance test, earthworms showed a clear preference for the S1.1 substrate, indicating this substrate to be most favorable for earthworms. Obviously, the combination of horse manure and leaves (components of the S1.1 substrate) is the most adequate for the earthworms. Comparison of additional properties of substrates (electrical conductivity, moisture, organic matter, and C/N ratio) did not reveal a relationship between observed avoidance results and measured properties.

The second and third sets of substrates were not subjected to composting, yet were used for vermicomposting right after mixing the components. In the second set of substrates, only one substrate (S2.2) proved to be unsuitable for vermicomposting, as substantial mortality was observed already after 48 h of exposure. Obviously, fresh horse manure (the only component of the S2.2 substrate) is not suitable for vermicomposting, yet additional components have to be added. Again, the remaining substrates (S2.1, S2.3, S2.4, S2.5, and S2.6) were further tested in an avoidance test. The results showed that earthworms had a preference for substrates S2.1, S2.4, and S2.6. When the composition of these substrates is compared, it is clearly visible that all substrates contained grape pomace. The grape pomace (the main solid by-product of the wine industry) has high concentrations of macro-nutrients and micro-nutrients that are easily available to plants due to high solubility in water and, therefore, its use in the vermicomposting process is recommended [36]. Additionally, it is considered that the grape pomace has a favorable effect on the growth and reproductivity of earthworms [37], and has antioxidant and antimicrobial properties [38]. In the present study, grape pomace also proved to be a good choice as a substrate in vermicomposting as the only component, as well as in combination with rock wool and fresh horse manure. Many studies have shown that rock wool is a very good medium for the growth of many ornamental plants such as chrysanthemums (*Chrysanthemum sp.*), but also various agricultural crops, such as tomatoes (*Solanaceae lycoperiscum*), peppers (*Capiscum annuum*), lettuce (*Lactuca sativa*), melon (*Cucumius melo*), and many others [38–40]. Therefore, this substrate is considered very suitable for growing seedlings, fruits, and vegetables. Obviously, grape pomace and rock wool can be used separately and in a mixture as substrates for vermicomposting. The usage of earthworms in the processing of these materials can contribute to obtaining a product with better characteristics.

In the third set of substrates, substantial mortality was observed in four substrates (S3.1, S3.4, S3.5, and S3.6). All these substrates contained fresh horse manure in a large proportion. Obviously, similar to substrate S2.2, fresh horse manure is not suitable for vermicomposting. If the fresh horse manure is used in vermicomposting, the medium could be phytotoxic. Phytotoxicity occurs due to various substances that are present in fresh horse manure, such as harmful trace elements, pathogens, ammonia, organic acids, phenol, salt, and others [41,42]. It is considered that horse manure is a favorable choice because it has significant amounts of carbon; however, precomposting should last 1–2 weeks [43]. Without precomposting, horse manure can cause high earthworm mortality and disable vermicomposting. The avoidance test was performed between two remaining substrates (S3.2 and S3.3), both containing grape pomace and sawdust, but S3.3 also had rock wool. Sawdust is considered a very good additive, given that it has a very high C:N ratio [44]. Although sawdust is dry, it removes unpleasant odors, so it can be used as an addition to various fertilizers [45]. Here, a combination of grape pomace and sawdust proved to be a good substrate for vermicomposting. However, the avoidance test showed a preference for substrate S3.3, indicating that the addition of rock wool to this combination additionally favors this substrate as a vermicomposting material.

The process of vermicomposting is influenced by various physicochemical factors, among which, the toxic heavy metals are of much concern since they may adversely affect earthworm activities and the overall vermicomposting process [46]. Even though the mortality and behavior assessment does show whether the substrate could be potentially used in vermicomposting, there could be still other effects of substrates on earthworms that could affect their efficiency in vermicomposting. Therefore, a preliminary biomarker assessment was performed and responses of selected biomarkers in earthworms exposed to investigated substrates were measured. Namely, activities of acetylcholinesterase (AChE), carboxylesterase (CES), and glutathione S-transferase (GST) were selected, as these biomarkers are one of the most common ones used in investigations of the impact of environmental pollutants on earthworms [47,48]. AChE is a biomarker of neurotoxicity; CES participates in phase I metabolism and GST is a phase II enzyme, and they are also involved in oxidative stress reactions. The obtained results showed that only substrates from the first set affected the activities of AChE, CES, and GST in earthworms. Namely, for all three enzymes, significant differences in activities between earthworms being exposed to different substrates were observed. Even though it is not possible to have any firm conclusions on the adverse effects of substrates on earthworms, the obtained results show that future investigations of substrate suitability should also include biochemical endpoints in order to fully address the substrate effects. Determination of the optimal substrate mixtures, in terms of optimal conditions for earthworms, will enable vermicomposting with optimal efficiency.

5. Conclusions

In order to achieve maximum efficiency in the vermicomposting process, it is advisable to carry out preliminary tests with previously selected substrates. In this way, the losses in terms of reduced efficiency and earthworm mortality during vermicomposting would be reduced. The avoidance test and measurement of biochemical parameters in earthworms proved to be adequate for that. For a more detailed insight into the effects of the substrates on earthworms in future studies, additional endpoints can be included—from the measurement of additional biochemical biomarkers to the assessment of biomass change and the reproduction of earthworms. Investigation of substrate suitability in this study showed that the most suitable substrates were leaves with horse manure and grape pomace alone and in combination with rock wool and sawdust. Therefore, the combinations of these components have the greatest potential to be used in vermicomposting. Products obtained by vermicomposting these mixtures should be evaluated in the production of organic fertilizers and substrates for growing seedlings.

Supplementary Materials: The following supporting information can be downloaded at https://www.mdpi.com/article/10.3390/agriculture12122128/s1. Table S1: Additional properties of investigated substrates and Table S2: Microorganisms added to the substrates.

Author Contributions: Conceptualization, S.E., M.V., A.M. and Z.L.; methodology, S.E. and M.V.; formal analysis, M.V.; investigation, S.E., J.B., L.S.K., S.K., J.J., F.N. and K.P.; resources, M.V. and Z.L.; writing—original draft preparation, S.E., M.V. and A.M.; writing—review and editing, all authors; supervision, M.V.; project administration, S.E. and M.V.; funding acquisition, Z.L. All authors have read and agreed to the published version of the manuscript.

Funding: The paper is the result of research within the project KK.01.1.1.04.0052 "Innovative production of organic fertilizers and substrates for growing seedlings" funded by the European Union under the Operational Programme of Competitiveness and Cohesion 2014–2020, the European Regional Development Fund.

Institutional Review Board Statement: Not applicable.

Data Availability Statement: Data are available upon request from the corresponding author.

Conflicts of Interest: The authors declare no conflict of interest.

References

1. Ose, A.; Andersone-Ozola, U.; Ievinsh, G. Substrate-dependent effect of vermicompost on yield and physiological indices of container-grown *Dracocephalum moldavica* Plants. *Agriculture* **2021**, *11*, 1231. [CrossRef]
2. Pierre-Louis, R.C.; Kader, M.A.; Desai, N.M.; John, E.H. Potentiality of vermicomposting in the South Pacific Island countries: A review. *Agriculture* **2021**, *11*, 876. [CrossRef]
3. Nogales, R.; Cifuentes, C.; Benitez, E. Vermicomposting of vinery wastes: A laboratory study. *J. Environ. Sci. Health Part B* **2005**, *40*, 659–673. [CrossRef]
4. Hussain, N.; Singh, A.; Saha, S.; Kumar, M.V.S.; Bhattacharya, P.; Bhattacharya, S.S. Excellent N-fixing and P-solubilizing traits in earthworm gut-izolated bacteria: A vermicompost based assessment with vegetable market waste and rice straw feed mixtures. *Bioresour. Technol.* **2016**, *222*, 165–174. [CrossRef] [PubMed]
5. Kouba, A.; Lunda, R.; Hlavac, D.; Kuklina, I.; Hamackova, J.; Randak, T.; Kozak, P.; Koubova, A.; Buric, M. Vermicompostin of sludge from reciculating aquaculture system using *Eisenia andrei*: Technological feasibility and quality assessment of end-products. *J. Clean. Prod.* **2018**, *177*, 665–673. [CrossRef]
6. Martin, M.; Eudoxie, M.M.G. Feedstock composition influences vermicomposting performance *Dichogaster annae* relative to Eudrilus eugeniae and *Perionyx excavatus*. *Environ. Sci. Pollut. Res.* **2018**, *25*, 17716–17725. [CrossRef]
7. Sharma, K.; Garg, V.K. Vermicomposting of waste: A Zero-Waste Apprroach for Waste Management. In *Suistanable Resource Recovery and Zero Waste Approaches*, 1st ed.; Taherzadeh, M., Bolton, K., Wong, J., Pandey, A., Eds.; Elsevier: St. Louis, MO, USA, 2019; pp. 133–164.
8. Sharma, S.; Kumar, A.; Singh, A.P.; Vasudevan, P. Earthworms and Vermitechnology—A Review. *Dyn. Soil Dyn. Plant* **2009**, *3*, 1–12.
9. Bhat, S.A.; Singh, J.; Vig, A.P. Earthworms as organic waste managers and biofertilizer producers. *Waste Biomass Valorization* **2017**, *9*, 1073–1086. [CrossRef]
10. Ramnarain, Y.I.; Ansari, A.A.; Ori, L. Vermicomposting of different organic materials using the epigeic earthworm *Eisenia foetida*. *Int. J. Recycl. Org. Waste Agric.* **2019**, *8*, 23–36. [CrossRef]
11. Vuković, A.; Velki, M.; Ečimović, S.; Vuković, R.; Čamagajevac, I.Š.; Lončarić, Z. Vermicomposting—Facts, benefits and knowledge gaps. *Agonomy* **2021**, *11*, 1952. [CrossRef]
12. Saranraj, P.; Stella, D. Vermicomposting and its importance in improvement of soil nutrients and agricultural crops. *Nov. Nat. Sci. Res.* **2012**, *1*, 14–23.
13. Manaig, E.M. Vermicomposting efficiency and quality of vermicompost with different bedding materials and worm food sources as substrate. *Res. J. Agric. For. Sci.* **2016**, *4*, 6063.
14. Hussain, N.; Abbasi, S.A. Efficacy of the vermicomposts of different organic wastes as "Clean" fertilizers: State-of-the-Art. *Sustainability* **2018**, *10*, 1205. [CrossRef]
15. Edwards, C.A.; Bohlen, P.J. Biology of Earthworms. In *Biology and Ecology of Earthworms*, 3rd ed.; Bohlen, P.J., Edwards, C.A., Eds.; Chapman & Hall: New York, NY, USA, 1996; 426p.
16. Sinha, R.K.; Soni, B.K.; Agarwal, S.; Shankar, B.; Hahn, G. Vermiculture for organic horticulture: Producing chemical-free, nutritive and health protective foods by earthworms. *J. Agric. Sci.* **2013**, *1*, 17–44. [CrossRef]
17. Ečimović, S.; Vrandečić, K.; Kujavec, M.; Žulj, M.; Ćosić, J.; Velki, M. Antifungal activity of earthworm coelomic fluid obtained from *Eisenia andrei*, *Dendrobaena veneta* and *Allobophora chlorotica* on six species of phytopathogenic fungi. *Environments* **2021**, *8*, 102. [CrossRef]
18. Edwards, C.A. Historical overview of vermicomposting. *Biocycle* **1995**, *36*, 56–58.

19. Ceritoğlu, M.; Şahin, S.; Erman, M. Effects of vermicompost on plant growth and soil structure. *Selcuk J. Agric. Food Sci.* **2018**, *32*, 607–615. [CrossRef]
20. Dominguez, J.; Edwards, C.A. Biology and ecology of earthworm species used for vermicomposting. In *Vermiculture Technology: Earthworms, Organic Wastes, and Environmental Management*; Edwards, C.A., Arancon, N.Q., Sherman, R.L., Eds.; CRC Press: Boca Raton, FL, USA, 2011.
21. Singh, J. Role of earthworm in sustainable agriculture. In *Sustainable Food Systems from Agriculture to Industry: Improving Production and Processing*; Academic Press: Cambridge, MA, USA, 2018; pp. 83–122.
22. Olee, M. Review: Vermicompost, its importance and benefit in agriculture. Estonian Crop Research Institute. *J. Agric. Sci.* **2019**, *30*, 93–98.
23. Thiruneelakandan, R.; Subbulakshmi, G. Vermicomposting: A superlative for soil, plant and environment. *Int. J. Innov. Res. Sci. Eng. Technol.* **2014**, *3*, 930–938.
24. Arancon, N.Q.; Edwards, C.E.; Atiyeh, R.M.; Metzger, J.D. Effects of vermicompost produced from food waste on the growth and yields of greenhouse peppers. *Bioresour. Technol.* **2004**, *93*, 139–144. [CrossRef]
25. Dominguez, J.; Gómez-Brandón, M. Vermicomposting: Composting with Earthworms to Recycle Organic Wastes. In *Management of Organic Waste*; Kumar, S., Ed.; Intech Open Science: London, UK, 2012; pp. 29–48.
26. Organization for Economic Cooperation and Development. *OECD Guidelines for Testing of Chemicals, Earthworm, Acute Toxicity Tests: Filter Paper Test and Artificial Soil Test*; Organization for Economic Cooperation and Development: Paris, France, 1984; Volume 207, pp. 1–9.
27. Bouché, M.B. Lombriciens de France. Ecologie et Systématique. In *Lombriciens de France*; INRA, Ed.; Institut National des Resherches Agriculturelles: Paris, France, 1972; pp. 72–78.
28. *ISO 17512-1:2008*; Soil Quality—Avoidance Test for Determining the Quality of Soils and Effects of Chemicals on Behavior—Part 1: Test with Earthworms (*Eisenia fetida* and *Eisenia andrei*). ISO: Geneva, Switzerland, 2008.
29. Smith, P.K.; Krohn, R.I.; Hermanson, G.T.; Mallia, A.K.; Gartner, F.H.; Provenzano, M.D.; Fujimoto, E.K.; Goeke, N.M.; Olson, B.J.; Klenk, D.C. Measurement of protein using bicinchoninic acid. *Anal. Biochem.* **1985**, *150*, 76–85. [CrossRef] [PubMed]
30. Ellman, G.L.; Courtney, K.D.; Andres, V.; Featherstone, R.M. A new and rapid colorimetric determination of acetylcholinesterase activity. *Biochem. Pharmacol.* **1961**, *7*, 88–95. [CrossRef] [PubMed]
31. Hosokawa, M.; Satoh, T. Measurement of carboxylesterase (CES) activities. *Curr. Protoc. Toxicol.* **2001**, *10*, 4–7. [CrossRef] [PubMed]
32. Habig, W.H.; Pabst, M.J.; Jakoby, W.B. Glutathione-S-transferase: The first enzymatic step in mercapturic acid formation. *J. Biol. Chem.* **1974**, *249*, 7130–7139. [CrossRef] [PubMed]
33. Haimi, J.; Hunta, V. Capacity of various organic residues to support adequate earthworm biomass for vermicomposting. *Biol. Fertil. Soils* **1986**, *2*, 23–27. [CrossRef]
34. Hund-Rinke, K.; Achazi, R.; Römbke, J.; Warnecke, D. Avoidance test with *Eisenia fetida* as indicator for the habitat function of soils: Results of a laboratory comparison test. *J. Soils Sediments* **2003**, *3*, 7–12. [CrossRef]
35. Lowe, C.; Butt, K.R.; Chenier, K.Y.-M. Assesment of avoidance behavior by earthworms (*Lumbricus rubellus* and *Octolasion cyaneum*) in linear pollution gradients. *Ecotoxicol. Environ. Saf.* **2016**, *124*, 324–328. [CrossRef] [PubMed]
36. Gomez–Brandon, M.; Lores, M.; Martinez–Cordeiro, H.; Dominguez, J. Effectiveness of vermicomposting for bioconversion of grape marc derived from red winemaking into a value-added product. *Environ. Sci. Pollut. Res.* **2019**, *27*, 33438–33445. [CrossRef]
37. Dominguez, J.; Martinez–Cordeiro, H.; Alvarez–Cases, M.; Lores, M. Vermicomposting grape marc yields high quality organic biofertiliser and bioactive polyphenols. *Waste Manag. Res.* **2014**, *32*, 1235–1240. [CrossRef]
38. Bussel, W.T.; Mckennie, S. Rockwool in horticulture, and its importance and sustainable use in New Zealand. *N. Z. J. Crop Hortic. Sci.* **2004**, *32*, 29–37. [CrossRef]
39. Hansen, M.; Gronborg, H.; Starkey, N.; Hansen, L. Alternative substrates for potted plants. *Acta Hortic.* **1993**, *342*, 191–196. [CrossRef]
40. Ellyard, R.K.; Ollerenshaw, P.J.; Hadobas, H.L. Growool as a cutting medium for Australian native plants. *Aust. Hortic.* **1985**, *83*, 26–27.
41. Hutchison, M.L.; Walters, L.D.; Avery, S.M.; Munro, F.; Moore, A. Analyses of livestock production, waste storage, and pathogen levels and prevalences in farm manures. *Appl. Environ. Microbiol.* **2005**, *71*, 1231–1236. [CrossRef] [PubMed]
42. Garcia-Sanchez, M.; Taušnerova, H.; Hanč, A.; Tlustoš, P. Stabilization of different starting materials through vermicomposting in a continuous-feeding system: Changes in chemical and biological parameters. *Waste Manag.* **2017**, *62*, 33–42. [CrossRef]
43. De la Vega, A. *Vermicomposting: The Future of Sustainable Agriculture and Organic Waste Management*; Lessons from the USA and Cuba; Royal Horticultural Society: London, UK, 2016; pp. 1–48.
44. Lončarić, Z.; Parađiković, N.; Popović, B.; Lončarić, R.; Kanisek, J. *Vegetables Fertilization, Organic Fertilizers and Composting*; Faculty of Agriculture, Josip Juraj Strossmayer University: Osijek, Croatia, 2015.
45. Lončarić, Z. Plodnost tala i gospodarenje organskim gnojivima. In Proceedings of the 11th Consultation of Cattle Breeders in the Republic of Croatia, Ivanić Grad, Croatia, 28–29 January 2016.
46. Kumar, S.; Sharma, V.; Bhoyar, R.V.; Bhattacharyya, J.K.; Chakrabarti, T. Effect of heavy metals on earthworm activities during vermicomposting municipal solid waste. *Water Environ. Res.* **2008**, *80*, 154–161. [CrossRef]

47. Velki, M.; Hackenberger, B.K. Different sensitivities of biomarker responses in two epigeic earthworm species after exposure to pyrethroid and organophosphate insecticides. *Arch. Environ. Contam. Toxicol.* **2013**, *65*, 498–509. [CrossRef]
48. Velki, M.; Ečimović, S. Important Issues in Ecotoxicological Investigations using Earthworms. In *Reviews of Environmental Contamination and Toxicology*; Springer International Publishing: Cham, Switzerland, 2016; pp. 157–184.

agriculture

MDPI

Article

Influence of Biochar and Animal Manures Application on Ammonia and Nitrate Concentrations in the Root and Shoot of Three Varieties of Turnips

George F. Antonious *, Eric T. Turley, Buddhi R. Gyawali and Angel C. Freeman

College of Agriculture, Community, and the Sciences, Division of Environmental Studies, Kentucky State University, Frankfort, KY 40601-2355, USA

* Correspondence: george.antonious@kysu.edu; Tel.: +1-502-597-6005; Fax: +1-502-597-5846

Abstract: Many investigators have focused on the impact of fertilizers on crop yield and ignored fertilizers impact on the plants composition. The impact of seven types of soil treatments (sewage sludge, horse manure, chicken manure, vermicompost, elemental organic fertilizer, inorganic fertilizer, and native soil) and similar seven treatments amended with biochar on the concentrations of NH_3 and NO_3 in the roots and shoots of three commercial varieties of turnips, *Brassica rapa* was investigated. The three varieties (Purple Top White Globe PTWG, Scarlet Queen Red SQR, and Tokyo Cross TC) varied in concentrations of NH_3 and NO_3 levels. High levels of NO_3 in edible plants is associated with harmful effects on human health, due to the risk of creation of carcinogenic N-nitroso compounds. NO_3 in SQR roots and shoots (edible greens) was greater than varieties PTWG and TC. The concentration of NH_3 averaged 20.2, 12.8, and 8.9 $\mu g\ g^{-1}$ fresh turnip roots, whereas NO_3 values averaged 107.6, 64.1, and 62.9 $\mu g\ g^{-1}$ fresh turnip roots in varieties SQR, PTWG, and TC, respectively. Regardless of soil amendment type, the concentration of NH_3 in the shoots (44.0 $\mu g\ g^{-1}$) was greater than the roots (15 $\mu g\ g^{-1}$). On the contrary, NO_3 was higher in the roots (89.4 $\mu g\ g^{-1}$) compared to the shoots (67.6 $\mu g\ g^{-1}$ fresh tissue). Overall, biochar added to vermicompost amended soil increased NH_3 by 73% compared to vermicompost not amended with biochar. Regarding acceptable daily intake (ADI) for NO_3, none of the three varieties analyzed constitute any NO_3 adverse effects on normal human intake. Similarly, consuming turnips grown in any of the animal manures tested do not represent any hazardous issues.

Keywords: vermicompost; chicken dung compost; horse dung compost; municipal sewage sludge; mineral inorganic fertilizer; organic fertilizer

1. Introduction

Nitrogen (N) supplied to plant roots as nitrate (NO_3) and ammonium ions (NH_4^+) is required in large amounts due to its greatest impact on plant growth [1], plant morphology, and nutrient composition [2]. Accordingly, N plays an important role in the yield and quality of growing plants. Most plants favor NO_3^-N since high concentration of NH_3-N is toxic during plant metabolism [3] and often recommended for application in small amounts after transplantation, due to irreversible alteration of the structure of the plant thylakoid membrane (the site of photochemical and electron transport reaction of oxygenic photosynthesis) [4]. Research results have indicated that the form of N supply has impact on photosynthesis, stomatal conductance and intercellular carbon dioxide (CO_2), but these results found not consistent among different plant species [5]. NH_3 in animal manures reacts with water to form ammonium ions (NH_4^+) that quickly binds to the negatively charged soil organic matter and clays. Nitrification in soil by *Nitrosomonas* and *Nitrobacter* bacteria oxidize NH_3 to NO_2 and NO_3. Plants uptake N from the soil in the form of NO_3, regardless of the form of N fertilizer applied, including animal manures. In humans, 5–10% of NO_3 is converted into the more toxic nitrite (NO_2) by salivary or gastrointestinal reduction. NO_2 can react with proteins in the body to form carcinogenic N-nitroso compounds, such as nitrosamines [6]. There are legal limits of NO_3 and NO_2 in food. They are hazardous chemicals that can accumulate in vegetables and fruits from application of fertilizers. Vegetables receive relatively high rates of N fertilizers, which adds to the problem of NO_3 poisoning due to vegetables ability to accumulate NO_3 at high levels [7]. Large-scale animal operating production systems yields huge amounts of manure rich in NO_3, which seeps into groundwater and accumulate in edible plants grown in animal manures amended soils. Therefore, keeping NO_3 concentrations below legal limits is a challenge for farmers and health authorities.

NO_3 acceptable daily intake (ADI) values of 0–3.7 mg NO_3 kg^{-1} body weight (BW) established by the Joint Expert Committee of the Food and Agriculture (JECFA) of the United Nations/World Health Organization (WHO) and the European Commissions of Scientific Committee on Food (SCF) [8]. Mensinga et al. [9] estimated 5–8% of the NO_3 from daily diets reduced to NO_2 by the microflora in the oral cavity. Assuming a drinking water consumption of 2 L per day and a daily consumption of 100 g of vegetables, the overall daily NO_3 consumption may easily range from 200 to 400 mg. The ADI of NO_3 estimated to be from 0–3.7 mg kg^{-1} BW, expressed as NO_3 or 277 mg NO_3 per person of 75 kg BW has been established [6].

Vegetables contain NO_3 at varying levels, ranging from 1 to 10,000 mg kg $^{-1}$ [10]. Vegetables can be classified according to their NO_3 content into very low (<200), low (200 to <500), middle (500 to <1000), high (1000 to <2500), and very high, (>2500 mg 100^{-1} g fresh weight), in which turnip has a middle range of 500 to <1000 mg 100^{-1} g fresh weight [11]. NO_3 accumulates in the mesophyll cells of the plant, since they exclusively transported among the plant tissue parts through the xylem [12]. In fact, the primary variables for NO_3 human intake includes the type of vegetables consumed, NO_3 levels in the type of vegetable consumed, and the amounts of vegetables consumed daily. The mean total NO_3 daily intake per person in Europe ranges between 50 and 140 mg and in the USA about 40–100 mg [13]. Toxic doses (with met hemoglobin formation as a criterion for toxicity) ranged from 33–350 mg NO_3- ion kg^{-1} BW [14] and human lethal doses of 67–833 mg NO_3^- ion kg^{-1} body weight (BW) reported. Consumption of one serving of a NO_3 rich food or supplement can exceed the World Health Organization acceptable daily intake for NO_3 (0–3.7 mg/kg body weight per day or 222 mg day^{-1} for a 60-kg adult).

Our hypothesis is that the use of soil amendments, such as animal manures that contain high levels of organic matter and nutrients is an inexpensive method to improve crop yield and soil quality. Reprocessing animal manures would reduce need of synthetic inorganic fertilizers and offer amendments useful for improving soil structure and nutrient composition at low-cost to small farmers. However, animal manures is a source of NH_3. NH_3 in animal manures reacts with water to form ammonium ions (NH_4^+) that quickly binds to the negatively charged soil organic matter and clays. In soil, NH_4^+ is transferred into nitrates (NO_3) that can also be absorbed by plants roots. NO_3 becomes a problem only if exceeded the allowable limits in food. We investigated the impact of animal manures used as organic fertilizers in agricultural production systems on the concentrations of NH_3 and NO_3 in three varieties of turnips, *Brassica rapa* grown in soil amended with animal manures and elemental organic and inorganic amendments on fresh weight basis. Quantification of NO_3 on a fresh-weight basis enables a better comparison of the NO_3 content of vegetables since most vegetables consumed fresh. Following food ingestion, bacteria in the mouth and gut convert NO_3 to NO_2 by salivary and gastrointestinal reduction, then NO_2 reacts with hemoglobin to produce met hemoglobin, which makes hemoglobin no longer able to carry oxygen [15]. Salehzadeh et al. [16] also reported that during various processes in the body, NO_3 usually converted to NO_2, which causes various diseases, such as blue baby syndrome and cancer. In fact, vegetable types and N fertilization influence NO_3 content in vegetables [8,17].

Because turnip can be grown in most locations and has short growing season (60 to 70 days) as a fall, winter, and early spring crop, turnips has a wide adaptation as a cash crop for limited-resource farmers. Vegetables are major source of NO_3 and NO_2 since they can reach values of 85% of the total intake in the human diet [18]. Monitoring the NO_3 content becomes an important indicator of the quality of plant products.

The literature review bare little information concerning the effect of animal manures and inorganic soil amendments on NO_3 concentrations in vegetable species and varieties in species. Researchers have focused on the crop yield and soil fertility after the incorporation of fertilizers with little attention to the plant internal composition. This study delivered indication of the low impact of animal manures on NO_3 levels in three commercial varieties of turnips and explained the danger of NO_3 accumulation in turnips and other edible plants.

Accordingly, the intend of this study was to identify turnip varieties and/or animal manures mixed and not mixed with biochar (a carbon-rich material produced during pyrolysis process of biomass) on the accumulation of NH_3 and NO_3 in turnip roots and edible shoots (turnip greens). The objectives were: (1)-assess the overall impact of six soil amendments: sewage sludge SS, horse manure HM, chicken manure CM, vermicompost Vermi, commercial organic fertilizer Nature Safe 10N-2P-8K) (Org), inorganic fertilizer (20N-20P-20K) (Inorg), and unamended native soil (UA native soil) on NH_3 and NO_3 concentrations in turnip roots and shoots. (2)-screen three varieties of turnips, *Brassica rapa* (Purple Top White Globe, Scarlet Queen Red and Tokyo Cross) for their accumulation of NH_3 and NO_3. (3)-investigate the impact of adding biochar to soil amendments on the concentration of NH_3 and NO_3 in fresh root and shoot of turnips grown under field conditions.

2. Material and Methods

The field study at the university of Kentucky Research Farm (Fayette County Lexington, Kentucky, USA Latitude: 37.976262, Longitude: −84.533334) included a randomized complete block design (RCBD) of 63 plots (3 turnip varieties × 7 treatments × 3 replicates) not treated with biochar and 63 plots treated with biochar. Native soil pre-experimental properties are: an average of 56% silt, 38% clay, and 6% sand, pH 6.8, CEC 14.7 meq 100 g^{-1}, OM 2.2%, Total N 0.18%, N-NO_3 20.7 mg L^{-1}, N-NH_4-N 5.7 mg L^{-1}, P 95.8 mg L^{-1}, K 336.2 mg L^{-1}, C 1091 mg L^{-1}, Cd 0.04 mg L^{-1}, Cu 1.9 mg L^{-1}, Zn, 1.98 mg L^{-1}, Pb 2.15 mg/L^{-1}, and Ni 0.66 mg L^{-1}.

Each of the 126 plots was 4 ft. (1.22 m) length and 3 ft. (0.91 m) width. The soil treatments included six soil amendments and unamended (UA) native soil used as control treatment. The six soil amendments were sewage sludge SS, horse manure HM, chicken manure CM, vermicomposting Vermi, commercial organic fertilizer (10N-2P-8K) Org, inorganic fertilizer (20N-20P-20K) Inorg. Each of the soil amendments used in this investigation was mixed with the native soil at 5% nitrogen (N) on dry weight basis to eliminate variations among soil treatments due to their variability in N content, since N fertilization has been identified as the major factor that influence NO_3 content of vegetables [8,17]. SS (5% N) purchased from the Metropolitan Sewer District in Louisville (KY, USA) and applied to native soil at 2241.7 kg hectare $^{-1}$. CM (1.1% N) obtained from the Department of Animal and Food Sciences, University of Kentucky (Lexington, KY, USA) and applied at 6592. 8 kg hectare^{-1}. HM (0.7% N) obtained from the Kentucky horse park (Lexington, KY, USA) was applied at 16,011.4 kg hectare^{-1}. Vermi (1.5% N) obtained from Worm Power (Montpelier, Vermont, USA) and applied at 9340.1 kg hectare^{-1}. Org (10% N) and Inorg (5% N) commercial fertilizers obtained from the Southern States Cooperative Stores (Lexington, KY, USA) and applied at 1120.9 and 560.4 kg hectare^{-1}, respectively (Table 1).

The three varieties of turnips, *Brassica rapa* were var. Purple Top White Globe (PTWG), var. Scarlet Queen Red (SQR), and var. Tokyo Cross (TC) (Figure 1). Prior to planting, each amendment added to native soil and rototilled to a depth of 15 cm (~0.5 ft.) top soil. Biochar (a carbon-rich material produced during pyrolysis and thermochemical decomposition of biomass), obtained from Wakefield Agricultural Carbon (Columbia, MO) was added at the rate of 10% (*w/w*). Properties of biochar used in this investigation were: total organic carbon 88%, total inorganic carbon 0.34%, surface area 366 m^2 g^{-1} dry, moisture 54%, temperature 200 °C, bulk density 480.6 kg m^{-3}, N 0.27%, P 2.06 mg kg^{-1}, K 280 mg kg^{-1}, Ca 1881 mg kg^{-1}, Cu 2.45 mg kg^{-1}, Mg 558 mg kg^{-1}, and Zn 2.09 mg kg^{-1}.

Table 1. Concentrations of NPK in animal manures, organic commercial fertilizer, and inorganic mineral fertilizer used for growing turnip, *Brassica rapa* (Fayette County, Kentucky, USA).

Soil Amendment	Nitrogen (% N)	Phosphorus (% P)	Potassium (% K)
Sewage Sludge	5.00	3.00	0.00
Chicken Manure	1.10	0.80	0.50
Horse Manure	0.70	0.30	0.60
Vermicompost	1.50	0.75	1.50
Organic Fertilizer	10.00	2.00	8.00
Inorganic Fertilizer	20.00	20.00	20.00
Amounts of Soil Amendments Added in kg hectare^{-1}			
Soil Amendment	**Nitrogen (N)**	**Phosphorus (P)**	**Potassium (K)**
Sewage Sludge	2241.7	1345.0	0.00
Chicken Manure	6592.8	4794.8	2996.7
Horse Manure	16,011.4	6861.9	13,724.0
Vermicompost	9340.1	4670.0	9340.1
Organic Fertilizer	1120.9	224.2	896.7
Inorganic Fertilizer	560.4	560.4	560.4

Soil amendments were applied to each treatment at 5% N. Determination of NPK was carried out using inductively coupled plasma (ICP) spectrometer.

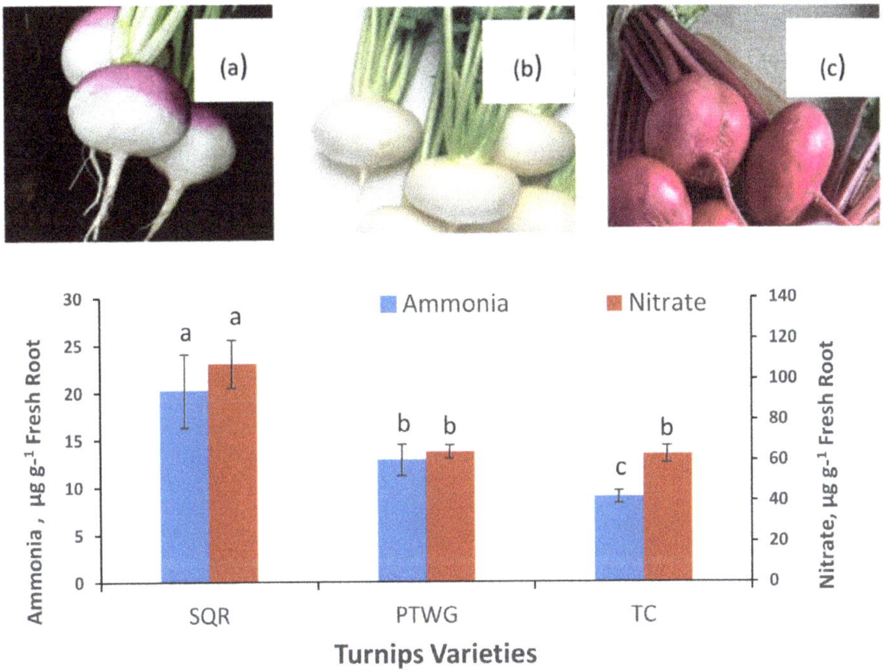

Figure 1. Variability in root morphology of three varieties of turnips (*Brassica rapa*): (**a**) Purple Top White Globe (PTWG), (**b**) Tokyo Cross (TC), and (**c**) Scarlet Queen Red (SQR) (upper photo) and NH_3 and NO_3 concentrations ± std. error in their roots (lower graph), regardless of soil treatments. Standard errors having different letter indicate significant differences ($p \leq 0.05$). Statistical comparisons carried-out among varieties using Duncan's multiple range test.

Seeds of turnip, *Brassica rapa* were planted in a freshly tilled soil at 45.7 cm in-row spacing, and the plants were drip irrigated as needed. Weeding and other agricultural operations carried out during the growing season regularly as needed. One month after planting, turnip plants were sprayed with the insecticides esfenvalerate (Asana XL) and Baythroide XL (β-cyfluthrin) three times during the growing season at the recommended rate of application [19]. At maturity (70 day old plants), three turnip varieties (PTWG, SQR, and TC) removed from the soil and their shoots (edible greens) and roots were separated using a sharp knife. Turnip greens are the dark leafy green tops that are edible and utilized in many cuisines. Five turnip plants randomly collected from each replicate (15 turnip plants from each treatment), and washed with deionized water for chemical analysis. Roots were cut vertically using a sharped knife into four quarters, one quarter from each root was cut into small cubes and a representative 100 g were selected for sample analysis. Similarly, the shoots (leaves and stems) were chopped using a kitchen shopper, extracted using 80% ethanol, and filtered using Whatman No. 1 filter paper. Quantification of NH_3 and NO_3 was carried out using a Fisher brand XL500 Benchtop Meter equipped with Orion High-Performance ammonia and nitrate electrodes (Fisher brand XL500 Benchtop Orion High-Performance ammonia and nitrate Electrodes) using the methods described by APHA [20].

Turnips roots, shoots, and plant weight were recorded. Concentrations of NH_3 and NO_3 in turnips roots and shoots were analyzed in each of the three turnip varieties) grown under the fourteen soil treatment. Data containing NH_3 and NO_3 in turnips root, shoot, and plant weight of each variety were statistically analyzed using one-way analysis of variance (ANOVA) (SAS Institute, 2016) [21] and the means were compared using Duncan's multiple range test.

3. Results

There were significant differences in NH_3 and NO_3 content among the three turnip varieties tested. Figure 1 shows that the concentrations of NH_3 averaged 20.2, 12.8, and 8.9 $\mu g\ g^{-1}$ fresh turnip roots, whereas NO_3 values averaged 107.6, 64.1, and 62.9 $\mu g\ g^{-1}$ fresh turnip roots in the three varieties (SQR, PTWG, and TC), respectively. These data revealed that variety SQR had significantly ($p \leq 0.05$) greater concentrations of NH_3 and NO_3 content compared to varieties PTWG and TC. Table 2 revealed a significant ($p \leq 0.001$) positive correlation (r = 0.57) between NH_3 and NO_3 content in variety PTWG grown in soil amended with biochar, while NH_3 and NO_3 were not significantly correlated in varieties TC and SQR. In addition, when biochar not added to soil, a significant negative correlation observed in variety TC. This negative correlation indicates that increasing the concentration of NH_3/NH_4^+ in variety TC is followed by low accumulation of NO_3, a needed attribute for increasing agricultural products human safety.

Table 2. Overall Pearson's correlation coefficients (r) and probability of significance (*P*) between ammonia and nitrates concentrations in turnip plants (root and shoot) grown in soil treated with biochar (A) and soil not treated with biochar (B).

(A)	PTWG	TC	SQR
Ammonia	r = 0.57	r = −0.17	r = −0.165
Nitrate	($p \leq 0.001$) *	($p = -0.2644$)	($p = 0.2956$)
(B)	PTWG	TC	SQR
Ammonia	r = −0.24	r = −0.51	r = −0.217
Nitrate	($p = 0.1397$)	($p = 0.006$) *	($p = 0.1723$)

Purple Top White Globe (PTWG), Tokyo Cross (TC), and Scarlet Queen Red (SQR); * indicates significant correlation ($p \leq 0.05$).

Regardless of turnip variety, results also revealed that Vermi, Inorg, CM, HM, and Org amended soil significantly increased NO_3 concentrations in turnip roots compared to the roots of plants grown in the unamended (UA) control treatment (Figure 2). In addition, soil amended with Vermi and SS increased NH_3 concentrations in turnip roots compared to other soil amendments and the control treatment (UA treatment). Although N content in the six amendments was applied at 5% N, Vermi and SS were superior in elevating NH_3 concentrations in turnip roots.

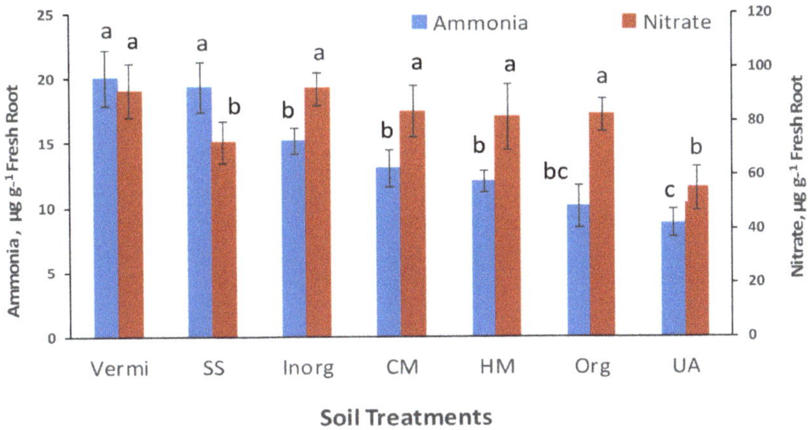

Figure 2. Concentrations of ammonia and nitrates ± std. error in turnips roots of plants grown under seven soil treatments, regardless of turnips variety. Vermi vermicompost, SS sewage sludge, Inorg inorganic fertilizer, CM chicken manure, HM horse manure, Org elemental organic fertilizer, and unamended (UA) native soil. Statistical analysis was carried-out using analysis of variance (ANOVA). Standard errors having different letter(s) indicate significant differences ($p \leq 0.05$) using Duncan's multiple range test.

As described earlier, the use of animal manures and mineral N fertilizers in agricultural production systems is a major source of ammonia (NH_3/ NH_4^+) emission [22], NH_3 emissions from animal manure used in agricultural production systems is generated by several physical, chemical, and biological processes [23]. Loss of NH_3/ NH_4^+ from manure are destructive, because they decrease the amount of manure N available for the crop and increase N contamination in groundwater through soil seepage (infiltration). Accordingly, variability among soil amendments, such as particle size, compaction, infiltration rate, moisture holding capacity, microbial activity, enzymes secretion, texture, pH, and other animal manures properties and composition are the main factors that control NH_3 emissions and NO_3 formation. However, regardless of amendment type used in this investigation, NO_3 in turnip greens of variety SQR had the highest concentration compared to PTWG and TC (Figure 3).

Regardless of turnip varieties, Figure 4 revealed that concentrations of NH_3 was greater in Vermi, SS, and Inorg treatments compared to CM, HM, Org, and UA control treatment, indicating that the addition of CM, HM, and Org fertilizer did not add NH_3 in turnips shoots of plants grown in unamended soil. In addition, all soil treatments increased the NO_3 content compared to the control (UA treatment). Results in Figure 5 revealed significant variability in NH_3 and NO_3 concentrations between turnip root and shoot. Concentrations of NH_3 averaged 44.2 and 14.9 µg g^{-1} fresh shoot and root tissue, respectively, whereas the concentrations of NO_3 averaged 67.6 and 89.4 µg g^{-1} fresh shoot and root tissue, respectively. These results of greater concentration of NO_3 in turnip roots compared to the shoot (edible greens) represents about 32% increase.

Regarding the impact of soil amendments, overall concentration of NH_3 in turnips root and shoot of plants grown in Vermi compost amended with biochar (VermiBio) was significantly ($p \leq 0.05$) greater (39.9 µg g^{-1} fresh tissue) compared to Vermi compost (Vermi) not amended with biochar (23.1 µg g^{-1} fresh tissue) (Figure 6). This significant increase revealed the positive impact of biochar (73.3% increase) on NH_3 concentration when biochar added to Vermi compost amended soil (VermiBio). Other than VermiBio, there was no impact of biochar addition on NH_3 concentrations in turnip plants before and after biochar addition. Figure 7 revealed that soil amended with inorganic fertilizer treated with biochar (InorgBio) significantly increased the concentration of NO_3 compared to biochar added to unamended control treatment (UABio). Other than InorgBio, no significant differences found in NO_3 concentrations among turnip plants grown in soil amendments treated with biochar and soil amendments not treated with biochar, regardless of turnip varieties.

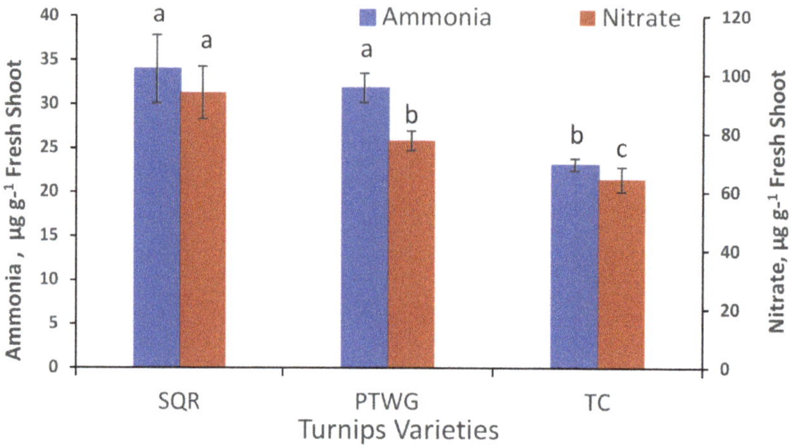

Figure 3. Variability in the concentrations of ammonia and nitrates ± std. error in the shoots of three varieties of turnips: Scarlet Queen Red (SQR), Purple Top White Globe (PTWG), and Tokyo Cross (TC), regardless of soil treatments. Statistical comparisons carried-out among varieties using analysis of variance (ANOVA). Standard errors having different letter indicate significant differences ($p \leq 0.05$) using Duncan's multiple range test for means comparison.

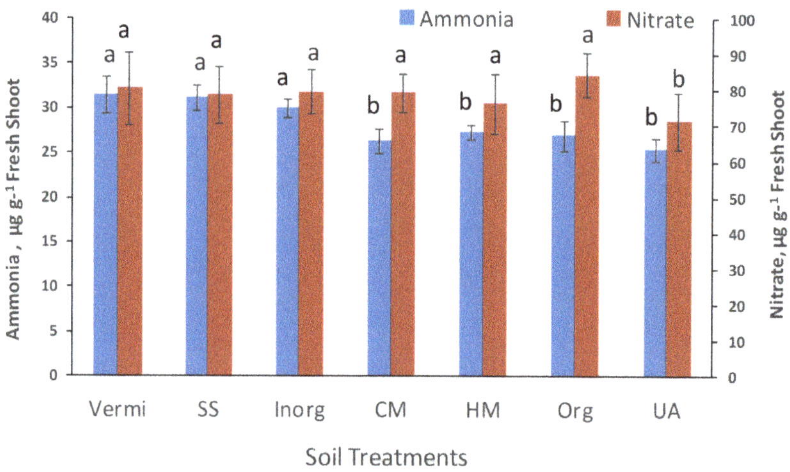

Figure 4. Concentrations of ammonia and nitrates ± std. error in turnips shoot of plants grown under seven soil treatments, regardless of turnips variety. Vermi vermicompost, SS sewage sludge, Inorg inorganic fertilizer, CM chicken manure, HM horse manure, Org elemental organic fertilizer, and unamended (UA) native soil. Statistical analysis was carried_out using analysis of variance (ANOVA). Standard errors having different letter indicate significant differences ($p \leq 0.05$) using Duncan's multiple range test for means comparison.

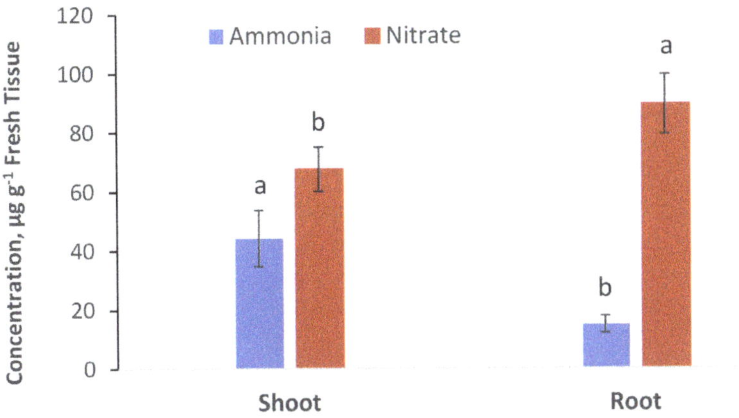

Figure 5. Overall concentrations of ammonia and nitrates in turnips shoot and root plants, regardless of turnips variety. Statistical comparisons were carried- out between the shoot and root using analysis of variance (ANOVA). Standard errors having different letter indicate significant differences ($p \leq 0.05$) using Duncan's multiple range test for mean comparisons.

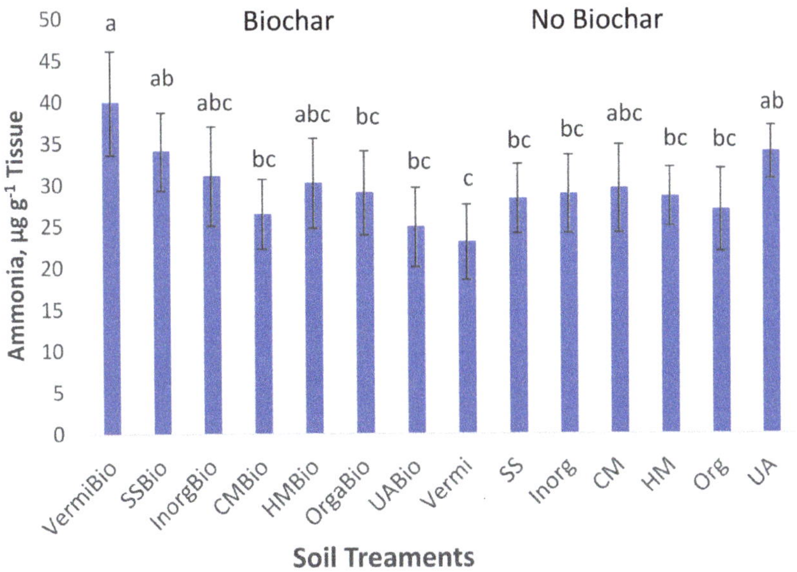

Figure 6. Concentrations of ammonia ± std. err in turnip plants grown under seven soil treatments not amended with biochar (vermicompost Vermi, sewage sludge SS, inorganic fertilizer Inorg, chicken manure, CM, horse manure HM, organic fertilizer Org, and unamended UA control), and seven soil treatments amended with biochar (VermiBio, SSBio, InorgBio, CMBio, HMBio, OrgBio, UABio), regardless of turnip varieties. Statistical analysis was carried out among 14 soil treatments using analysis of variance (ANOVA). Standard errors having different letter(s) indicate significant differences ($p \leq 0.05$) using Duncan's multiple range test for means comparison.

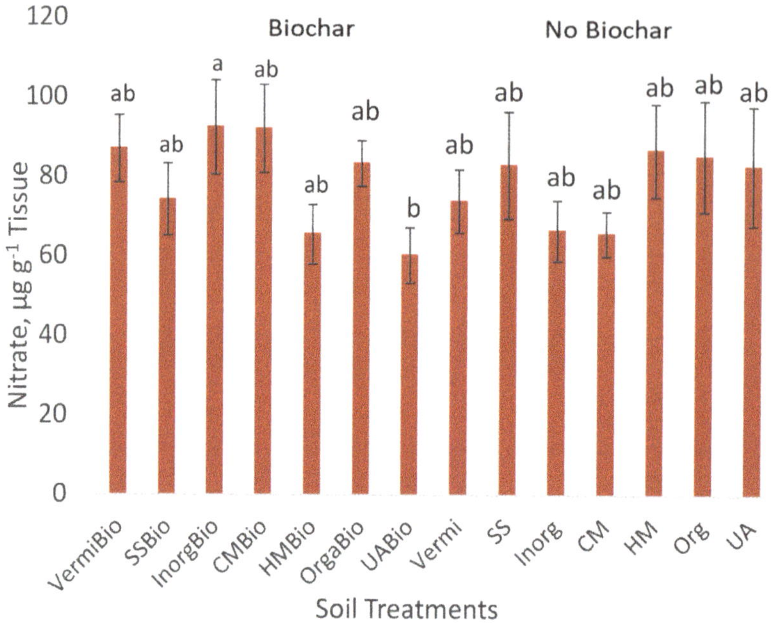

Figure 7. Concentrations of nitrate in turnip plants grown in seven soil treatments not amended with biochar (vermicompost Vermi, sewage sludge SS, inorganic fertilizer Inorg, chicken manure, CM, horse manure HM, organic fertilizer Org, and unamended UA control soil), and seven soil treatments amended with biochar (VermiBio, SSBio, InorgBio, CMBio, HMBio, OrgBio, UABio), regardless of turnip varieties. Statistical analysis was carried out among 14 soil treatments using analysis of variance (ANOVA). Standard errors having different letter(s) indicate significant differences ($p \leq 0.05$) using Duncan's multiple range test for means comparison.

Antonious et al. [24] reported that NO_3 concentrations in Vermi was significantly greater compared to other animal manures. In addition, urease activity (the enzyme that breakdown urea forming NH_4^+ and CO_2) also was greater in Vermi. NO_3 toxic doses due to methaemoglobin formation (exposure of hemoglobin to a variety of highly reactive oxygen free radicals produced during normal cell metabolism), ranged from 33–350 mg NO_3^- ion kg^{-1} body weight (BW) have been reported by Speijers [14]. The oral lethal dose to humans was estimated to vary from 33 to 250 mg NO_2^- ion kg^{-1} BW. Doses of 1 to 8.3 mg NO_2^- ion kg^{-1} BW, gave rise to induction of methemoglobinemia in which the hemoglobin iron (Fe) oxidized and cannot reversibly bind oxygen [14]. Salehzadeh et al. [16] reported that a person with an average weight of 70 kg should not consume more than 255.5 mg of NO_3 daily. Boink and Speijers [6] reported that the acceptable daily intake (ADI) for NO_3 is assigned as 0–3.7 mg kg^{-1} body weight (BW) or 277 mg NO_3 per person of 75 kg average weight.

Table 3 revealed a significant ($p \leq -0.69$) negative correlation (r = 0.0014) between NH_3 and NO_3 content in turnip plants grown in soil amended with municipal SS mixed with biochar. A significant ($p \leq 0.0014$) negative correlation (r = −0.69) between NH_3 and NO_3 content in turnip plants grown in soil amended with vermicompost not mixed with biochar was also obtained, while correlations between NH_3 and NO_3 in turnips grown in other soil amendments were not significantly correlated. This negative correlation indicates that increasing the concentration of NH_3/NH_4^+ is followed by low accumulation of NO_3, which is a needed attribute for increasing human safety.

Table 3. Overall Pearson's correlation coefficients (r) and probability of significance (*P*) between ammonia and nitrates concentrations in turnip plants (root and shoot) grown in soil treatments treated with biochar (A) and soil treatments not treated with biochar (B).

(A)	Vermi	SS	Inorg	CM	HM	UA
Ammonia Nitrate	r= −0.15 (p = 0.54)	r = −0.69 (p = 0.0014) *	r = 0.18 (p = 0.47)	r = 34 (p = 0.158)	r = 0.08 (p = 0.74	r = 0.14 (p = 0.57)
(B)	Vermi	SS	Inorg	CM	HM	UA
Ammonia Nitrate	r= −0.694 (p = 0.0014) *	r = 0.43 (p = 0.069)	r = 0.0015 (p = −0.69)	r = −0.31) (p = 0.214)	r = 0.05 (p = 0.84)	r = 0.13 (p = 0.58)

Vermi vermicompost, SS sewage sludge, Inorg inorganic fertilizer, CM chicken manure. HM horse manure, Org elemental organic fertilizer, and UA unamended control treatment. * indicates significant correlation (p < 0.05).

4. Discussion

Vegetables and animal manures contain ammonia (NH_4^+) and nitrate (NO_3^-) ions that constitute a potential health hazard to consumers. NH_4^+ and NO_3^- are natural constituents of vegetables and fruits. The toxic effects of NO_3 are due to its endogenous conversion to nitrite (NO_2^- in saliva and human gastrointestinal tract. NO_3 toxicity among vegetable consumers and growers interested in growing turnips in animal manures amended soils have received increased attention, which resulted in several investigations on the dietary exposure to these compounds. Investigators reported that about 5% of the dietary NO_3^- reduced to NO_2^- in saliva and the gastrointestinal tract and this number might reach 20% for individuals with a high rate of conversion [25].

Hmelak and Cencic [11] reported that NO_3 extensively distributed in nature and different concentrations of NO_3 are detected in soil, water, and food, but ingestion and exposure to NO_3 is mainly from vegetables and water. A moderate reduction in plants yield caused by NH_4 + stress could be prevented by the application of nitrification inhibitors, such as 2-chloro-6-(trichloromethyl) pyridine (nitrapyrin), dicyandiamide (DCD), and 3,4-dimethylepyrazole phosphate (DMPP) with NH_4+ fertilizers or organic fertilizers which makes high concentrations of NH_4 + stable in the soil for several weeks. On the other hand, the plant cell has several strategies to keep NH_4+ levels under control either by NH_4+ efflux to the plant rhizosphere area, or by storing NH_4+ in the cell vacuole, or by NH_4+ incorporation into organic compounds [26]. The positive effects of low concentrations of NO_3 and NO_2 presented by Parvizishad et al. [27] could have a protective effect on the cardiovascular system, blood pressure regulation, and maintaining homeostasis (stability) of vessels. The authors [27] discussed the different opinions about the allowable concentrations of NO_3 and NO_2 in food and water. They concluded that these compounds have beneficial and adverse effects on human health, and encouraged the need of more research to make proper judgments about setting the standards concentration in food and drinking water.

Due to the danger of the potential high NO_3 levels in food, several studies in different institutions and countries around the world monitored and established their own regulations for the control of NO_3 contaminations of vegetables [28]. Wu et al. [29] found that inappropriate vegetable cultivation methods, such as excessive use of nitrogen fertilizers could certainly cause extreme NO_3 accumulation in leafy vegetables. In addition, the unsuitable vegetable cooking processes can trouble their NO_3 balance and potential NO_2 safety risk. The authors detected a decrease in NO_3 content and the rapid increase in NO_2 content during 12–24 h of storage due to the effect of microorganisms in the storage environment. Kyriacou et al. [30] also reported that due to the abuse of chemical fertilizers and unreasonable planting methods, the NO_3 content of intensively planted vegetables tend to reach excessively high NO_3 levels. Salehzadeh [16] investigated the impact of cooking vegetables on NO_3 concentrations in relation to health risks of NO_3 in vegetables. They found that NO_3 concentration in leafy vegetables was higher than root and fruit vegetables and these values were higher in autumn than in spring growing seasons. The results of their study revealed that cooking reduced NO_3 levels and lowers the health risk of eating raw vegetables. Recently Xu et al. [31] studied the effect of N fertilizer rates on NH_3 oxidizing archaea (AOA) and NH_3 oxidizing bacteria (AOB) community. They found that the major phyla of AOA and AOB were *Thaumarchaeota* and *Proteobacteria*, respectively. They also conducted a correlation analysis between AOA and AOB abundance and found that AOA abundance showed significantly positive correlations with soil pH, and negative correlation with soil NH_4-N, NO_3-N, whereas AOB abundance positively correlated with soil NO_3-N, but negatively correlated with soil pH. Li et al. [32] applied a model to clarify the factors affecting loss of NH_3 and NO_3 from greenhouse vegetables. They found that drip irrigation amplified NH_3 volatilization and reduced NO_3 leaching by 20 kg N ha^{-1} and 75 kg N ha^{-1}, respectively, whereas combining drip irrigation with lessening N application by 50%, significantly decreased greenhouse gas emission without any sacrifice in vegetable yield.

Investigators have found an amplified risk of thyroid cancers with developed NO_3/NO_2 intake [33,34] and high NO_3 absorption is associated with bigger risk of cancers in urinary bladder [35]. High levels of NO_3 may also decrease the nutritional value of consumed vegetables as it affects carotenoid, vitamins A and B degradation [36].

We monitored the concentration of NH_4^+ and NO_3^- in roots and shoots of three field grown varieties of turnips, *Brassica rapa* (Purple Top White Globe PTWG, Scarlet Queen Red SQR, and Tokyo Cross TC) grown under soil mixed with six types of soil amendments mixed and not mixed with biochar. In this study, we found greater NO_3^-level in turnip roots (89.4 µg g^{-1} fresh tissue) compared to the shoot (67.6 µg g^{-1} fresh tissue). On the contrary, NH_4^+ level was greater in the shoot (44.2 fresh tissue) compared to the roots (15 µg g^{-1} fresh tissue), regardless of turnips variety. Several factors, such as consumption of other vegetables and amount consumed per person and per day might contribute to NO_3 toxicity in human diet. Overall, there was a significant increase (73%) in NH_3 concentration in turnips plants (root and shoot) grown in Vermi compost amended with biochar compared to Vermi not amended with biochar. Soil amended with inorganic fertilizer treated with biochar significantly increased the concentration of NO_3 by 35%, compared to biochar control treatment. The observed variability among turnip varieties and soil amendments applied in this investigation might be attributed to variability within turnip varieties and activity of amendments' hydrolyzing enzymes, such as nitrate reductase, urease, as well as the type of fertilizer applied. In fact, consuming turnips is not the only source of NO_3 intake. Other sources of NO_3 such as drinking water and other foodstuff determine the actual health risk associated with NO_3 ingestion.

Our future objectives will focus on monitoring the impact of animal manures on the activity of nitrate reductase (the enzyme that reduce the conversion of NO_3^- to nitrite (NO_2^-) and urease (the enzyme that hydrolyze urea to NH_4^+ and CO_2 in field-grown vegetables and fruit species in relation to the allowable NO_3^- intake. We will also monitor the mobility of NH_4^+ and NO_3^- from animal manures amended soil into runoff and seepage water following natural rainfall events under field conditions that influence the quality of natural water resources.

5. Conclusions

The average content of NO_3 detected in each of the three turnip varieties tested in this investigation (Figure 1) indicated that the concentration of NO_3 in variety SQR is greatest (108 µg g^{-1} fresh root tissue) compared to the PTWG and TC varieties (64 and 62 mg kg^{-1} fresh root tissue, respectively). Therefore, a person with an average weight of 75 kg consuming 100 g of variety SQR would have 10.8 mg NO_3 in his diet and 0.14 mg kg^{-1} NO_3 per BW. These values would be 0.09 and 0.09 mg kg^{-1} BW for consuming turnip varieties PTWG and TC, respectively. Biochar increased the concentration of NH_3 in turnip plants grown vermicompost amended soil by 73% compared to vermicompost not amended with biochar. Other than that, there was no impact of biochar addition on NH_3 concentrations in turnip plants before and after biochar addition, regardless of turnip varieties (Figure 6). Addition of biochar to inorganic fertilizer (InorgBio) significantly increased NO_3 concentration by 35% compared to unamended native soil treated with biochar (Figure 7).

Based on our investigation, the assigned ADI for NO_3 range of 0–3.7 mg kg^{-1} BW is acceptable and none of the three varieties tested could cause any NO_3 adverse effects on average human consumption. Similarly, consuming turnip shoot (edible greens) grown in any of the animal manures amended soil do not represent any hazardous issues. We concluded that the quantity and quality of elemental fertilizers as well as animal manures applied in agricultural production systems are crucial aspects that regulate the concentration of NH_4^+ and NO_3^- absorbed from soil amendments into edible plants.

The average total intake of NO_3^- level per person in USA ranges between 40–100 mg day^{-1} [9]. In this study, we found that NO_3 contents never exceeded the EU limit concentration of 200 mg kg^{-1} BW. The average total intake of NO_3 per person in Europe ranges between 50 and 140 mg day^{-1} and in the USA about 40–100 mg day^{-1} [9]. According to the European Union Legislation [37] on food contaminants, concentrations of NO_3 in the root and shoot of turnips in each of three varieties tested or among the soil amendments treated with biochar or no-biochar, NO_3 concentrations never exceeded the permitted limits in turnips.

The application of animal manure as organic fertilizer has important properties that cannot be obtained from synthetic inorganic fertilizers. Microorganisms in animal manures facilitate the slow release of the three main plant nutrients, N, P, and K from soil organic matter, reducing their offsite mobility to natural water resources and eutrophication. The literature review revealed a lack of information regarding the impact of organic and inorganic amendments on NO_3 concentrations in vegetable species and varieties within species. Investigators have focused on the plant yield and soil physical and chemical characteristics following the incorporation of fertilizers with very little information on the plant internal composition. We provided evidence of the low impact of animal manures (a great source of N) on NO_3 levels in three varieties of turnips that reduce or eliminate the danger of NO_3 accumulation in fresh turnips roots and shoots.

Author Contributions: G.F.A. designed the study, conducted the laboratory analysis, and wrote the manuscript. E.T.T. and A.C.F. worked on the field study and project administration. B.R.G. participated in reviewing the manuscript. All authors have read and agreed to the published version of the manuscript.

Funding: This investigation was supported by two joint grants from the United States Department of Agriculture, National Institute of Food and Agriculture (USDA/NIFA) under agreement #KYX-10-18-65P Accession 1017900 & National Science Foundation/Historically Black College Universities (NSF/HBCU-UP) grant award #HRD 2011917 to Kentucky State University.

Institutional Review Board Statement: Not applicable (the study did not involve animals or humans).

Informed Consent Statement: Not applicable (the study did not involve humans).

Data Availability Statement: The study did not report any data.

Acknowledgments: The authors recognize Steven Diver and his farm team for maintaining the tomato field plots.

Conflicts of Interest: The authors affirm that there is no conflict of interest.

References

1. Gallegos-Vázquez, C.; Olivares-Sáenz, E.; Vázquez-Alvarado, R.; Zavala-García, F. Nitrate and ammonium uptake by cactus pear in hydroponics. *Terra Latinoam.* **2000**, *18*, 133–139. (In Spanish)
2. Bernstein, N.; Ioffe, M. Effects of supplied nitrogen form and quantity on growth and postharvest quality of Ranunculus asiaticus flowers. *Hortic. Sci.* **2005**, *40*, 1879–1886. [CrossRef]
3. Barker, V.A.; Mills, A.H. Ammonium and nitrate nutrition of horticultural crops. *Hortic. Rev.* **1980**, *2*, 395–423.
4. Simonne, E.; Simonne, A.; Wells, L. Nitrogen source affects crunchiness, but not lettuce yield. *J. Plant Nutr.* **2001**, *24*, 743–751. [CrossRef]

5. Zhou, Y.; Zhang, X.; Wang, J.; Cui, X.; Xia, K.; Shi, K.; Yu, J.Q. Effects of nitrogen form on growth, CO_2 assimilation, chlorophyll fluorescence and photosynthetic electron allocation in cucumber and rice plants. *J. Zhejiang Univ. Sci. B* **2011**, *12*, 126–134. [CrossRef]
6. Boink, A.; Speijers, G. Health effects of nitrates and nitrites, a review. *Acta Hortic.* **2001**, *563*, 29–36. [CrossRef]
7. Lorenz, O.A. Potential nitrate levels in edible plant parts. *Soil–Plant–Nitrogen Relatsh.* **1978**, *2*, 201–219. [CrossRef]
8. Santamaria, P. Nitrate in vegetables: Toxicity, content, intake and EC regulation (review). *J. Sci. Food Agric.* **2006**, *86*, 10–17. [CrossRef]
9. Mensinga, T.T.; Speijers, G.J.A.; Meulenbelt, J. Health implications of exposure to environmental nitrogenous compounds. *Toxicol. Rev.* **2003**, *22*, 41–51. [CrossRef]
10. Correia, M.; Barroso, A.; Barroso, F.M.; Soares, D.; Oliveira, M.B.P.P.; Delerue-Matos, C. Contribution of different vegetable types to exogenous nitrate and nitrite exposure. *Food Chem.* **2010**, *120*, 960–966. [CrossRef]
11. Hmelak Gorenjak, A.; Cencič, A. Nitrate in vegetables and their impact on human health: A review. *Acta Aliment.* **2013**, *42*, 158–172. [CrossRef]
12. Blom-Zandstra, M. Nitrate accumulation in vegetables and its relationship to quality. *Ann. Appl. Biol.* **1989**, *115*, 553–561. [CrossRef]
13. Tamme, T.; Reinik, M.; Roasto, M.; Juhkam, K.; Tenno, T.; Kiis, A. Nitrates and nitrites in vegetables and vegetable-based products and their intakes by the Estonian population. *Food Addit. Contam.* **2006**, *23*, 355–361. [CrossRef]
14. Speijers, G.J.A. Nitrate. In *Toxicological Evaluation of Certain Food Additives and Contaminants in Food*; WHO Food Additives Series; WHO: Geneva, Switzerland, 1996; Volume 35, pp. 325–360.
15. Katan, M.B. Nitrate in foods: Harmful or healthy? *Am. J. Clin. Nutr.* **2009**, *90*, 11–12. [CrossRef]
16. Salehzadeh, H.; Maleki, A.; Rezaee, R.; Shahmoradi, B.; Ponnet, K. The nitrate content of fresh and cooked vegetables and their health- related risks. *PLoS ONE* **2020**, *15*, e0227551. [CrossRef]
17. Gruda, N. Impact of environmental factors on product quality of greenhouse vegetables for fresh consumption. *Crit. Rev. Plant Sci.* **2005**, *24*, 227–247. [CrossRef]
18. Gangolli, S.D.; van den Brandt, P.A.; Feron, V.J.; Janzowsky, C.; Koeman, J.H.; Speijers, G.J. Nitrate, nitrite and N-nitroso compounds. *Environ. Toxicol. Pharmacol.* **1994**, *292*, 5–32. [CrossRef]
19. Rudolph, R.; Pfeufer, E.; Bessin, R.; Wright, S.; Strang, J. Vegetable Production Guide for Commercial Growers, University of Kentucky College of Agriculture, Food and Environment Cooperative Extension Service 2020. Available online: http://www2.ca.uky.edu/agcomm/pubs/id/id36/id36.pdf (accessed on 1 October 2022).
20. American Public Health Association (APHA). *Standard Methods of Examination of Water and Wastewater*, 21st ed.; Eaton, A.D., Clesceri, L.S., Rice, E.W., Greenberg, A.E., Eds.; Port City: Baltimore, MD, USA, 2005.
21. SAS Institute Inc. *SAS/STAT Guide*, Version 9.4. Campus Drive. SAS Institute Inc.: Cary, NC, USA, 2016; p. 27513.
22. Rostami1, M.; Monaco, S.; Sacco, D.; Grignani, C.; Dinuccio, E. Comparison of ammonia emissions from animal wastes and chemical fertilizers after application in the soil. *Int. J. Recycl. Org. Waste Agric.* **2015**, *4*, 127–134. [CrossRef]
23. Monaco, S.; Sacco, D.; Pelissetti, S.; Dinuccio, E.; Balsari, P.; Rostami, M.; Grignani, C. Laboratory assessment of ammonia emission after soil application of treated and untreated manures. *J. Agric. Sci.* **2012**, *150*, 65–73. [CrossRef]
24. Antonious, G.F.; Dawood, M.H.; Turley, E.T.; Paxton, R.B. Yield and quality of lettuce, pumpkin, and watermelon varieties grown under five-soil management practices. *Int. J. Appl. Agric. Sci.* **2021**, *7*, 57–65. [CrossRef]
25. Thomson, B.M.; Nokes, C.J.; Cressey, P.J. Intake and risk assessment of nitrate and nitrite from New Zealand foods and drinking water. *Food Addit. Contam.* **2007**, *24*, 113–121. [CrossRef]
26. Marino, D.; Moran, J.F. Can ammonium stress be positive for plant performance? *Front. Plant Sci.* **2019**, *10*, 1103. [CrossRef]
27. Parvizishad, M.; Dalvand, A.; Hossein, M.A.; Goodarzi, F. A Review of Adverse Effects and Benefits of Nitrate and Nitrite in Drinking Water and Food on Human Health. *Health Scope* **2017**, *6*, e14164. [CrossRef]
28. Ranasinghe, R.A.S.N.; Marapana, R.A.U.J. Nitrate and nitrite content of vegetables: A review. *J. Pharmacogn. Phytochem.* **2018**, *7*, 322–328.
29. Wu, S.; Liu, Y.; Cui, X.; Zhang, Q.; Wang, Y.; Cao, L.; Luo, X.; Xiong, J.; Ruan, R. Assessment of Potential Nitrite Safety Risk of Leafy Vegetables after Domestic Cooking. *Foods* **2021**, *10*, 2953. [CrossRef] [PubMed]
30. Kyriacou, M.C.; Soteriou, G.A.; Colla, G.; Rouphael, Y. The Occurrence of Nitrate and Nitrite in Mediterranean Fresh Salad Vegetables and Its Modulation by Preharvest Practices and Postharvest Conditions. *Food Chem.* **2019**, *285*, 468–477. [CrossRef] [PubMed]
31. Xu, A.; Li, L.; Xie, J.; Gopala Krishnan, S.; Zhang, R.; Luo, Z.; Cai, L.; Liu, C.; Wang, L.; Anwar, S.; et al. Changes in Ammonia-Oxidizing Archaea and Bacterial Communities and Soil Nitrogen Dynamics in Response to Long-Term Nitrogen Fertilization. *Int. J. Environ. Res. Public Health* **2022**, *19*, 2732. [CrossRef] [PubMed]
32. Li, G.; Xie, H.; Zhang, J.; Li, H. WHCNS-Veg Modelling of N_2O, NH_3 and NO_3 dynamics in a vegetable production system under different fertilization and irrigation regimes. *Atmosphere* **2022**, *13*, 1289. [CrossRef]
33. Ward, M.H.; Kilfoy, B.A.; Weyer, P.J.; Anderson, K.E.; Folsom, A.R.; Cerhan, J. Nitrate intake and the risk of thyroid cancer and thyroid disease. *Epidemiology* **2010**, *21*, 389–395. [CrossRef]
34. Kilfoy, B.A.; Zhang, Y.; Park, Y.; Holford, T.R.; Schatzkin, A.; Hollenbeck, A.; Ward, M.H. Dietary nitrate and nitrite and the risk of thyroid cancer in the NIH-AARP Diet and Health Study. *Int. J. Cancer* **2011**, *129*, 160–172. [CrossRef]
35. Michaud, D.S.; Mysliwiec, P.A.; Aldoori, W.; Willett, W.C.; Giovannucci, E. Peptic Ulcer Disease and the Risk of Bladder Cancer in a Prospective Study of Male Health Professionals. *Cancer Epidemiol. Biomark.* **2004**, *13*, 250–254. [CrossRef]

36. Raczuk, J.; Wadas, W.; Głozak, K. Nitrates and nitrites in selected vegetables purchased at supermarkets in Siedlce, Poland. *Rocz. Państwowego Zakładu Hig.* **2014**, *65*, 15–20.
37. European Commission. Commission Regulation (EC) No. 466/2001 of 8 March 2001 Setting Maximum Levels for Certain Contaminants in Foodstuffs. 2001. Available online: https://scirp.org/reference/referencespapers.aspx?referenceid=878481 (accessed on 1 October 2022).

MDPI AG
Grosspeteranlage 5
4052 Basel
Switzerland
Tel.: +41 61 683 77 34

Agriculture Editorial Office
E-mail: agriculture@mdpi.com
www.mdpi.com/journal/agriculture

www.ingramcontent.com/pod-product-compliance
Lightning Source LLC
Chambersburg PA
CBHW062050220926
51850CB00010B/2718